Encyclopedia of Herbicides: Properties and Synthesis

Volume II

Encyclopedia of Herbicides: Properties and Synthesis

Volume II

Edited by **Molly Ismay**

New York

Published by Callisto Reference,
106 Park Avenue, Suite 200,
New York, NY 10016, USA
www.callistoreference.com

Encyclopedia of Herbicides: Properties and Synthesis
Volume II
Edited by Molly Ismay

International Standard Book Number: 978-1-63239-256-5 (Hardback)

Printed in the United States of America.

Contents

Permissions

List of Contributors

Preface

This book covers topics of synthesis and various properties of herbicides in an in-depth manner. It outlines the detailed studies of various synthetic pathways of specific herbicides and the physical and chemical features of other synthesized herbicides. The objective of this book is to showcase several characteristics and features of herbicides, the physical and chemical properties of specific types of herbicides, and their effects on physical and chemical features of soil and micro-flora. In addition, an assessment of the extent of contamination in soils as well as crops by herbicides has been elaborated along with a research on the performance and photochemistry of herbicides and the impact of excess herbicides in soils and field crops.

This book is the end result of constructive efforts and intensive research done by experts in this field. The aim of this book is to enlighten the readers with recent information in this area of research. The information provided in this profound book would serve as a valuable reference to students and researchers in this field.

At the end, I would like to thank all the authors for devoting their precious time and providing their valuable contribution to this book. I would also like to express my gratitude to my fellow colleagues who encouraged me throughout the process.

Editor

Synthesis and Properties

Benzoxazolinone Detoxification and Degradation – A Molecule´s Journey

Margot Schulz[1], Dieter Sicker[2], František Baluška[3], Tina Sablofski[1,3],
Heinrich W. Scherer[4] and Felix Martin Ritter[1]

[1]*University of Bonn, Institute of Molecular Physiology
and Biotechnology of Plants (IMBIO)*
[2]*University of Leipzig, Institute of Organic Chemistry*
[3]*Institute of Cellular and Molecular Botany (IZMB)*
[4]*Institut für Nutzpflanzenwissenschaften und Ressourcenschutz(INRES)
Plant Nutrition
Germany*

1. Introduction

Benzoxazinoids are important secondary products of maize, several other Poaceae and a few dicotyledonous species belonging to the Acanthaceae, Lamiaceae, Scrophulariaceae and Ranuculaceae. The synthesis which was investigated in maize by the group of Gierl and Frey starts with the conversion of indole-3-glycerol phosphate to indole. The following steps involve four cytochrome P450 dependent monooxygenases (*BX2-BX5*) that convert indole to benzoxazinone by incorporation of oxygen. Glucosylation at the 2-position of DIBOA results in DIBOA-glucoside, an intermediate of the final product DIMBOA-glucoside (Frey et al., 1997; Glawischnig et al., 1999; von Rad et al., 2001; Jonczyk et al., 2008; Schuhlehner et al., 2008). Whereas the benzoxazinoid acetal glucosides are stable under neutral conditions, the aglycones with the 2,4-dihydroxy-2*H*-1,4-benzoxazin-3(4*H*)-one skeleton underlay a degradation by ring contraction and release of formic acid which yields the benzoxazolinones BOA or MBOA (Sicker et al., 2000; Sicker & Schulz, 2002). These derivatives are more stable and can be detected in the soil of rye or wheat fields over a period of several weeks until they are absorbed by other plants or they are converted by microorganisms. The release of benzoxazinoids into the environment and their final degradation are cornerstones within the lifetime of these molecules. In between, a complex set of (re)-modulations and conversions take place due to the activities of a variety of organisms, such as higher plants, fungi and bacteria. Our contribution will give an impression of shuttles between those organisms that end up in the degradation of phenoxazinone(s) as the final conversion products with a limited life time but will also present several reactions of maize to the treatment with benzoxazolinone BOA.

Investigations of weed specific and of benzoxazinoid producing crops specific reactions, reactions of microorganisms, effects on the biodiversity of soil organisms and the elucidation of degradation processes are unequivocally necessary before bioherbicides can be used.

2. Functions of benzoxazinoids

Benzoxazinone glucosides are stored in the vacuole until the tissue is damaged, for example by herbivores, and hydrolysis of the sugar moiety by ß-glucosidases takes place. The highly bioactive aglycones can be released into the environment also by root exudation or by plant residue degradation (Barnes & Putnam, 1987). The mutagenic benzoxazinoids are electrophilic compounds that interact with proteins, intercalate with nucleic acids and are deleterious for many cellular structures and activities (Frey et al., 2009; Sicker & Schulz, 2002). In maize, DIMBOA may have an additional endogenous function. Recently Frebortova et al. (2010) discussed a possible role in cytokinin degradation. Oxidative cleavage of DIMBOA led to coniferron, an electron acceptor of cytokinin dehydrogenase. However, benzoxazinoids have first of all an outstanding role as chemical weapons against other organisms (Niemeyer 2009). Aside of their insecticidal, fungicidal and bactericidal properties, benzoxazinoids are phytotoxic to susceptible plants. Often observed reactions are an inhibited germination but particularly the reduction of seedlings growth. Therefore, the compounds could play an important role in sustainable agricultural systems for natural weed and pest control in innovative agricultural systems.

3. Factors that influence benzoxazinone accumulation

The amount of benzoxazinoids varies highly with plant age, organ and cultivar. Investigated rye cultivars differ in the total benzoxazinoid amounts from 250 to 1800 μg g^{-1} dry tissue in young plants to about 100 μg g^{-1} or less in old plants (Reberg-Horton et al., 2005; Rice et al., 2005; Zasada et al., 2007). In rye cultivars used as mulches by Tabaglio et al. (2008), the content ranges from 177 to 545 μg g^{-1}. High differences in the concentrations among rye cultivars are also reported by Burgos et al. (1999). Water stress conditions and high temperatures increase the content of DIMBOA and DIBOA (Gianoli &Niemeyer 1997; Richardson & Bacon 1993). Nitrogen fertilization has a significant influence on the benzoxazinoid content (Gavazzi et al., 2010). In maize, we found a 3-4 fold higher benzoxazinone accumulation under sulfur deficiency conditions compared to the control plants which were cultivated under optimal nutrient supply (Fig. 1).

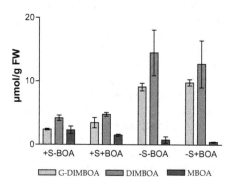

Fig. 1. Influence of sulfur on the benzoxazinoid accumulation in maize roots. Three week old plants were used for the incubation with 0.5 mM BOA (40 ml=20 μmol/g FW). Plant cultivation, BOA incubation, extraction and analyses were performed as described in Knop et al. (2007) and Sicker et al. (2001). N =5.

The reason for the stress induced benzoxazinoid accumulation is unclear since the biosynthesis of the compounds is developmentally regulated (Frey et al., 2009). Recently Ahmad et al. (2011) found an increased apoplastic accumulation of DIMBOA-glucoside, DIMBOA and HDMBOA-glucoside in maize leaves during defined stages of infestation with *Rhopalosiphum padi* and *Setosphaeria turtica*. Thus, the translocation of benzoxazinoids out of the cell may be an important step of a process which can lead to an increased stress tolerance and biocidal defense.

4. Effects of benzoxazolinone in maize – increase of glutathione transferase actvity and glutathione levels

Several groups investigated the mode of action of benzoxazolinones (Baerson et al., 2005; Batish et al., 2006; Sanchez-Moreiras et al., 2010, 2011; Singh et al., 2005) on *Lactuca sativa*, *Arabidopsis thaliana* or *Phaseolus aureus*. In plants BOA induces oxidative stress, membrane damage and lipid peroxidation. A prolonged exposure to high BOA concentrations (45 µmol, *Arabidopsis thaliana*) up to 8 days led to a decline in photosynthetic efficiency, induced senescence and death. At sub lethal concentrations, *A. thaliana* reacts with a strong alteration of the gene expression pattern, which comprises about 1% of the total genome. Burgos et al. (2004) found reduced densities of ribosomes, dictyosomes and mitochondria together with a lower amount of starch granules in roots of cucumber seedlings after treatment with BOA or DIBOA. These authors assume that BOA and DIMBOA induces changes in cellular ultrastructure, reduces root growth by disrupting lipid metabolism, by a decreased protein synthesis, and by a reduced transport or reduction of secretory capabilities.

Although maize roots are relatively resistant to BOA (Knop et al., 2007), their physiology is affected when exposed for 24 hours to levels considered to be non-toxic (500 µM and lower). As indicated by the marker compound malondialdehyde (MDA) lipid peroxidation is one of the earliest effects in roots of 6 to 7 days old maize seedlings. An increase is already observed after 1 min and a maximum between 5 to 40 min (Fig. 2). Subsequently, the MDA amount drops below the control value. This indicates the fast activation of mechanisms that counteract cellular damage, also an important action to avoid autotoxicity. During the next hours the level of GST activity is slightly increased (17-30% compared to +S -BOA conditions) in BOA incubated root tips of plants cultivated under optimal sulfur supply. At the same time the major detoxification product glucoside carbamate starts to accumulate (see below). Root tips from -S-plants have only about 40% of the GST activity found in +S-plants. The activity increases up to 50% during the course of incubation, but the presence of BOA has no influence (Fig. 3). Thus, -S-plants have deficits in providing GSTs that have a function in stress reactions.

The soluble plant glutathione transferases are categorized in defined classes: Θ (GSTF), T (GSTU), Φ (GSTT), Z (GSTZ), Λ (GSTL), dehydroascorbate reductase and tetrachlorohydroquinone dehalogenase like enzymes. Phi and tau class enzymes have a well known function in herbicide detoxification. GSTs respond to many processes that induce ROS production. Up regulating of GST gene expression can be triggered by herbicide safeners (Riechers et al., 2010). GSTs could have as well a function in the detoxification of endogenous substrates (Dixon et al., 2010). Because of the lack of *in vivo* accumulating natural glutathionylated products, Dixon et al. (2010) postulate unstable reaction products, which may decay or which are immediately transformed to other products by metabolic channeling. GSTs may have also non-catalytic functions as transporters of unstable GS-conjugates, which can be generated spontaneously via radical formation of an acceptor molecule in presence of glutathione. In *Arabidopsis*, BOA induces the up regulation of several GST genes (Baerson et al., 2005). If these GSTs are involved in BOA detoxification pathways is unclear since glutathione conjugates have not yet been found in *Arabidopsis*, maize or other plants. However, it cannot be excluded that GSTs have a role in the transport of unstable intermediates of BOA detoxification products. This question is currently under investigation.

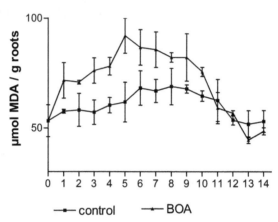

Fig. 2. Course of malondialdehyde (MDA) production in maize roots of 6-d-old seedlings. 0: control; 1-5: intervals of measurements 1 min; 6-10: intervals 15 min; 11: 90 min, 12: 3h, 13: 3h, 14: 4h after start of the incubation. Seedlings were grown and incubated as described in Schulz & Wieland (1999). Samples were prepared and MDA was determined according the method of Wong et al. (2001). $N = 5$.

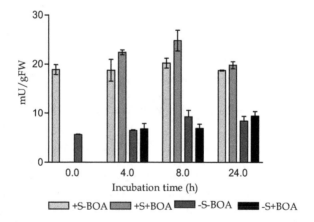

Fig. 3. GST activity in root tips of aeroponically cultured maize plants (greenhouse condition) according the method of Habig & Jacoby (1981). Sulfur deficiency was induced as described in Knop et al. (2007). $N = 3$.

The tripeptide glutathione (GSH) is the major thiol inter alia in plants and a substrate for the GSTs. The multiple functions of GSH in organisms include important roles in redox-homeostatic buffering, in cellular signaling, root development, sulfur assimilation, in defense and stress reactions and in the detoxification of xenobiotics (see review article of Noctor et al., 2011).

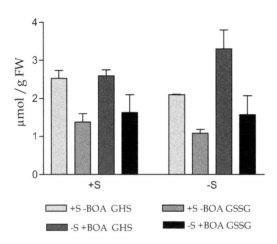

Fig. 4. Reduced and oxidized glutathione in root tips of aeroponically cultured maize plants (greenhouse). The plant material was extracted as described by Gamczarka (2005); measurement of glutathione was done according to Teare et al. (1993). $N = 3$.

The glutathione concentration measured in +S-root tips is similar to the values published by Kocsy et al. (2000). In -S-root tips, the total GHS content is decreased. However, we found slightly increased content of the total GHS in the root tips of –S-maize plants after 24 h exposure to BOA (Fig. 4). The content of GSSH in all samples indicates the induction of some oxidative stress due to the incubation conditions (see also Fig. 2), but there is a tendency of higher GSSG contents in BOA incubated plant material which indicates oxidative stress. Our results are conform to the findings that herbicides can increase the content of glutathione and the activity of associated enzymes such as glutathione reductase in herbicide tolerant plants (Kocsy et al., 2000). The response of –S-maize glutathione content to exogenously applied BOA is obviously very similar to that observed with certain herbicides. Glutathione synthetase gene induction under stress conditions is known (Hruz et al., 2008). An induction of glutathione synthesis even under -S-conditions would involve a change in the priority of the sulfur use in sulfur deficient plants and the mobilization of sulfur from sulfur containing molecules in the root tip. The exact role of glutathione in BOA detoxification is still under investigation. At present it is already unambiguous that the sulfur availability is important in the plant´s coping with BOA (see below).

5. Actin cytoskeleton, cytoarchitecture, and auxin transport

Although maize root growth is not inhibited significantly by BOA, we have scored subtle but relevant effects on the actin cytoskeleton and root apex cytoarchitecture which are stronger exhibited in –S-plants. In cells of the meristem and transition zone, nuclei are affected in their typical central position and shifted laterally and/or axially (Fig. 5, 6). Similar effects were reported in maize root cells having affected their actin cytoskeleton due to impacts of the actin polymerization inhibitors or in mutant of maize *lilliputian* having aberrant actin cytoskeleton, irregular cell files and root anatomy (Baluška et al., 2001a, 2001b). Importantly, the actin cytoskeleton under the plasma membrane (Fig. 8), especially at the synaptic cell-cell adhesion domains is affected (Baluška et al., 2005). These domains are depleted in their abundant F-actin (Baluška et al., 1997) whereas there are over-polymerized F-actin foci assembled around nuclei of BOA-exposed root cells, shifted out of cellular centres (Fig. 6-8). Similar impacts on the cytoarchitecture were reported in maize root cells exposed to vesicular secretion inhibitor refeldin A (Baluška & Hlavacka 2005), as well as to mastoparan which is affecting phosphoinositide signalling (Baluška et al., 2001c), and to auxin transport inhibitors (Schlicht et al., 2006). Unique BOA-induced effect on the actin cytoskeleton of the transition zone cells is the prominent local assembly of F-actin patches at corners of cross-walls (plant synapses) which resemble published data of the *brk1* mutant line of *Arabidopsis* (Fig. 7C, D) in this chapter and Figure 8 in Dyachok et al., 2008). Interestingly, the BRK1 protein localizes to the cross-wall corner sites of *Arabidopsis* root apices showing aberrant actin organization both in the root cells of the *brk1* mutant and in the BOA-exposed root cells (Fig. 4). BRK1 is a component of the evolutionary conserved SCARE complex that acts as F-actin nucleator and BOA might directly target the SCARE complex. This would be a very attractive scenario and it should be tested in future. BOA is also known to inhibit activity of the PM H^+-ATPase of root cells (Friebe et al., 1997) and it is of interest to note that the actin cytoskeleton is controlling permeability of the plant plasma membrane (Hohenberger et al., 2011). Interestingly, maize mutants *lrt1* and *rum1*, and especially the *lrt1/rum1* double mutant, which are affected in the polar auxin transport, showed similar F-actin depletion at the synaptic cell-cell adhesion domains and shifted nuclei (Schlicht et al., 2006). In general, all polar auxin inhibitors resemble BOA in affecting the actin cytoskeleton especially at the transition zone of the root apex which is the most active zone with respect of F-actin rearrangements (Baluška et al., 1997, 2001a, 2001b, 2001d), the polar auxin transport (Baluška et al., 2010; Mancuso et al., 2005, 2007).

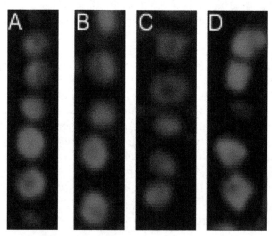

Fig. 5. Shifted nuclei (visualized with DAPI) in root apex cell files. **A:** +S-BOA, **B:** -S-BOA, **C:** +S+BOA, **D:** -S+BOA.

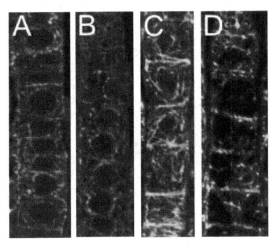

Fig. 6. Overview of the actin cytoskeleton in maize root apex. **A**: +S-BOA, **B**: -S-BOA, **C**: +S+BOA, **D**: -S+BOA.

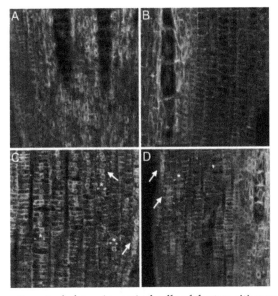

Fig. 7. Details of the actin cytoskeleton in cortical cells of the transition zone. Note the aberrant over-polymerization of F-actin in cortex cells of the BOA-exposed roots. **A**: +S-BOA, **B**: -S-BOA, **C**: +S+BOA, **D**: -S+BOA

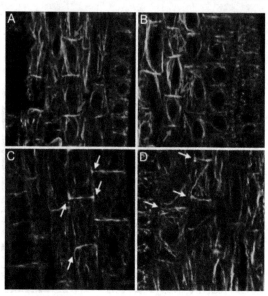

Fig. 8. Details of the actin cytoskeleton in cells of the transition zone. Red arrows indicate abundant F-actin at the cross walls (plant synapses) of pericycle and endodermis cells. White arrows indicate depleted F-actin at the cross walls (plant synapses) of pericycle/ endodermis cells of the BOA-exposed roots. Note the over-polymerization of F-actin in the cell corners. **A**: +S-BOA, **B**: -S-BOA, **C**: +S+BOA, **D**: -S+BOA.

For the experiments, apical root segments (~7mm) encompassing the major growth zones were excised, fixed with 3.7% formaldehyde, and embedded in the Steedman's wax. Ribbons of 7-mm sections were dewaxed and incubated with a mouse anti-actin monoclonal antibody, clone C4 from ICN Pharmaceuticals (Costa Mesa, CA, USA) diluted 1:200 and a rabbit maize polyclonal anti-actin (gift of Chris Staiger, Purdue University, USA) diluted 1:100 in PBS buffer. After rinsing with PBS buffer and incubation with secondary antibodies, sections were mounted under a coverslip and examined in the confocal microscope Olympus Fluoview 1000.

Similarly to *Arabidopsis* roots, the actin cytoskeleton is affected by auxin transport inhibitors in similar manner as we have reported here for the maize root cells (Rahman et al., 2007; Dhonukshe et al., 2008). Importantly in this respect, BOA is known to act as anti-auxin and to block lateral root formation (Anai et al., 1996; Baerson et al., 2005; Burgos et al., 2004; Hoshi-Sakoda et al., 1994) a process well-known to be based on auxin transport in both monocots and dicots roots (Hochholdinger & Zimmermann, 2008; Peret et al., 2009). Finally, the root apex transition zone emerges as specific target of allelochemicals, particularly this unique zone of the root apex (Baluška et al., 2010). Future studies should focus on these effects of BOA on the polar auxin transport and on other processes and activities characteristic for the root apex transition zone (Baluška et al., 2010).

6. Detoxification of benzoxazolinones in higher plants

Plants react to BOA in a species- and dosage dependent manner. Generally, members of the Poaceae were found to be less sensitive to the compounds than dicotyledoneous species, although there are exceptions. Moreover, ecotypes (for instance, *Chenopodium album* ecotypes or *Portulaca oleracea* garden forms) and varieties can differ in their accumulation and detoxification activities (Schulz unpublished). One important reason for the different sensitivity is the better developed ability of most Poaceae to reduce the toxicity benzoxazolinone(s), in comparison to dicots. Deleterious effects on the biochemistry, physiology and cell biology are therefore limited in good detoxifiers (see effects on maize described above). Interestingly, Macias et al. (2005) found an almost 100% inhibition of *Allium cepa* and *Lycopersicon esculentum* root growth whereas *Triticum aestivum* root growth was inhibited to 50 %, when low concentrations of DIMBOA-glucoside (5 µmol = 5 ml of a 1mM solution used as the highest concentration) were applied to 10 or 25 seeds in Petri dishes. It is generally known that the glucosides of benzoxazinones are much less toxic than the aglycons, but they obtained similar results with DIBOA. The growth of *Lepidium sativum* was stimulated to about 20%.

Almost all investigated higher plant species detoxify benzoxazolinone (BOA) via 6-hydroxylation and subsequent O-glucosylation (Tab. 1). *Portulaca oleracea* and a few other related species produce BOA-5-O-glucoside as a byproduct (Hofmann et al., 2006). Monocots perform, mainly or at least to a considerable portion, glucoside carbamate (Schulz et al., 2006; Schulz & Wieland, 1999; Sicker et al., 2000, 2004; Wieland et al., 1998). In contrast to BOA-6-OH and its glucoside, glucoside carbamate is not toxic up to concentrations of 1 mM and is therefore a most suitable detoxification product. First found in maize, glucoside carbamate is subsequently modified by malonylation or by addition of a second glucose molecule yielding gentiobioside carbamate (Hofmann et al., 2006). BOA-6-O-glucoside is, however, the major detoxification product when maize or other seedlings are incubated with MBOA. Glucoside methoxycarbamate occurred only in maize as a minor compound when the incubation was extended to more than 48 h. Identified stable detoxification products are illustrated in figure 9. The accumulation of BOA-6-OH is a good marker for a high sensitivity to BOA (for example *Vicia faba*). This hydroxylation product is twice as toxic as BOA and causes necrosis in the root tips within 24 h.

The BOA-detoxification process in maize roots starts with the production of BOA-6-O-glucoside. However, after 3 to 6 h glucoside carbamate accumulation is initiated. About 10 h after incubation start, this compound becomes the major detoxification product, whereas BOA-6-O-glucoside does not further accumulate although the glucosyltransferase activity responsible for glucosylation of BOA-6-OH is still abundant (Schulz et al., 2008). The increasing accumulation of gentiobioside carbamate and malonyl-glucoside carbamate 18 to 20 h after start of the incubation is a late event in the detoxification process. Avoidance of BOA uptake can be another strategy to escape the harmful effects of BOA.

In a recent study, we found a significant reduction of redroot pigweed (*Amaranthus retroflexus* L.) and common purslane (*Portulaca oleracea* L.), whereas common lambsquarters (*Chenopodium album* L.) and velvetleaf (*Abutilon theophrasti* Medicus) were moderately or not suppressed, respectively (Gavazzi et al., 2010; Tabaglio et al., 2008).

One possibility to explain the different reactions of the four weeds could be differences in the detoxification activities or accumulation characteristics that minimize the harmful effects of rye allelochemicals (BOA and related compounds). This affects a direct correlation between the benzoxazinoid content of rye mulch used in the study and weed suppression. The four warm season weeds exhibit remarkable differences in their detoxification behavior with a high correlation to the sensitivities of the weeds previously observed in experiments with rye mulch under greenhouse conditions. These studies demonstrate for the first time that detoxification processes are important for the survival of adapted weeds in environments enriched with benzoxazinoids, such as maize, wheat or rye fields (Schulz et al., submitted). Moreover, nutrients together with stress conditions have an influence on the detoxification processes. For instance, sulfur deficiency in combination with herbicide treatment can lead to a breakdown of the BOA detoxification process in maize (Knop et al., 2007). Optimal sulfur supply seems to be an emerging factor to guarantee well functioning of detoxification pathways. This is particularly important since sulfur deficiency is increasing in many areas of the word (Scherer, 2001, 2009).

Family	Species	A	B	C	D	E	F
Poaceae	*Avena sativa*	xx	xx				
	Avena fatua	xxx	xx				
	Digitaria sanguinalis	xxx	x				
	Lolium perenne	xx	xxx				
	Hordeum vulgare	xx	xxx				
	Triticum aestivum	x	xxx				
	Secale cereale	x	xxx				
	Zea mays	(x)	xxx		xx	xxx	
Portulacaceae	*Portulaca oleracea* cv Gelber	xx	xx		x	x	xx
Chenopodiaceae	*Chenopodium album (ecotypel)*	xx	x	x			
Brassicaceae	*Arabidopsis thaliana*	xxxx	x				
	Rhaphanus sativus	xxx	x				
	Diplotaxis tenuifolia	xx		x			
Amaranthaceae	*Amaranthus albus*	xxx					
	Amaranthus retroflexus	xx	xx		x	x	
Ranunculaceae	*Consolida orientalis*	xxx	x				
	Consolida regalis	xxx	x				
Apiaceae	*Coriandrum sativum*	xxx	xx				
	Daucus carota	xxx	x	x			
Asteraceae	*Galinsoga ciliate*	xxx		x			
	Helianthus annuus	xxx					
Fabaceae	*Vicia faba*	xx		xx			

Table 1. Some plant species (6-10 days old seedlings) and their major BOA detoxification products after 24 h of incubation with 0.5 mM BOA (40 ml / g FW). Maize: compounds present after 48 h are considered. Major compound: xxx; xx: minor compound: x; traces (x). A: BOA-6-O-glucoside; B: glucoside carbamate; C: BOA-6-OH; D: gentiobioside carbamate; E: malonylglucoside carbamate; F: BOA-5-O-glucoside.

There are also some hints that the ecobiochemical potential of species to detoxify benzoxazolinone drives the membership to certain plant associations (Schulz & Wieland, 1999).

A portion of the detoxification products are released again by root exudation (Sicker et al., 2002). When BOA incubated maize plants are transferred to tap water, BOA-6-O-glucoside and glucoside carbamate can be identified in the water. After several days, the compounds cannot be detected anymore in the soluble fraction prepared from the plants. A similar result is obtained with *Galinsoga ciliata* and *Coriandrum sativum*, indicating that exudation of soluble detoxification products is a more general phenomenon. The exuded products can get in contact with endophyts and microorganisms of the rhizosphere.

7. Microbial degradation products and fate of exuded plant degradation products

Many fungi are known to be sensitive to benzoxazinones and benzoxazolinones. However, some are able to detoxify the compounds (Fig. 10). Species of *Fusarium* have been investigated for their growth in presences of benzoxazinone (Friebe et al. 1998; Glenn et al. 2001). Eleven of 29 *Fusarium* species had some tolerance to BOA, the most tolerant species was *F. verticillioides* with only one sensitive strain of the 56 ones tested (Glenn et al., 2001). The first step in the degradation of benzoxazolinone-2(3*H*)-one (BOA) is a hydrolysis yielding 2-aminophenol. This step is performed by bacteria as well, also by seed born ones (Bacon et al. 2007; Burdziak et al., 2001). 2-Aminophenol is not stable but is spontaneously dimerized to 2-amino-3*H*-phenoxazin-3-one (APO) or it can be captured by several fungi which convert the compound to N-(2-hydroxyphenyl)malonamic acid (oHPMA) and 2-acetamidophenol (AAP) (Carter et al., 1999; Friebe et al., 1998; Glenn et al., 2001, 2002, 2003). Several endophytic fungi (*Plectosporium tabacinum, Gliocladium cibotii, Chaetosphaeria sp., Fusarium sambucinum*) from *Aphelandra tetragona* are described to produce 2-amino-(3*H*)-phenoxazinone derivatives when incubated with benzoxazinones (Baumeler et al., 2000; Zikmundova et al., 2002a, 2002b). *Fusarium verticillioides*, an endophytic fungus of maize, did not convert benzoxazolinone to any known microbial degradation product when sterile grown maize seedlings were inoculated with the fungus whereas the seedlings produced their known detoxification products since gentiobioside carbamate and glucoside carbamate could be detected in the medium. APO, AAP and oHPMA can have effects on plant growth. Absorbed traces of AAP and oHPMA stimulated maize radicle growth; traces of AAP stimulated that of cress. Phenoxazinone inhibited the growth of cress radicles at concentrations higher than 500 µM, whereas maize radicles were hardly affected (Knop et al., 2007).

In another study (Schulz et al., unpublished), the growth of some representative fungi was monitored over a period of 10 days in presence of BOA and APO. Generally, BOA was always less toxic than APO. The ability to grow in presence of BOA is influenced by the availability of nutrients. Several species changed the sensitivity to BOA, when BOA had to be used as N-source.

Once released into the soil, the plant and microbial detoxification products can be degraded by fungi. All compounds are finally converted to phenoxazinone(s): The degradation work of *Botrytris cinerea* (B.cin), *Drechslera tuberose* (D.tub), *Fusarium heterosporum* (F.het), *F. verticillioides* , *F. oxysporum* (F.oxy), *F. culmorum* (F.cul) , *F. solanum* (F.sol*)*, *Trichoderma viride* (T.vir) is presented in Fig. 11.

In the media of *Fusarium verticillioides* and *Drechslera tuberosa*, some benzoxazolinone-2(3*H*)-one (BOA) is present after the incubation with glucoside carbamate, the medium of the other fungi contained only traces of BOA. This indicates an opening of the carbamate heterocycle followed by the release of glucose. *Botryis cinerea* has only a rather limited ability to degrade

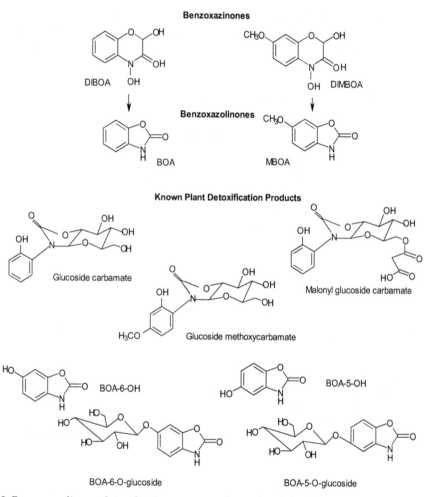

Fig. 9. Benzoxazolinone detoxification compounds produced by plants.

the compound, while *Paecilomyces farinosus* is unable to convert it. The behavior is, however, highly dependent on the different strains of a given species.

In the media of all of the species able to degrade glucoside carbamate a new, hitherto unknown intermediate occurred. The new compound was isolated for structural analysis. The ¹H spectrum showed signals for an *ortho*-substituted phenyl ring and well resolved signals in the sugar region with all couplings, too. The complete assignment was made by use of H,H-COSY, HMQC and HMBC. The latter technique was decided to prove that the hydrolytic ring opening of the oxazolinone precursor 1-(2-hydroxyphenylamino)-1-deoxy-ß-glucoside 1,2-carbamate (glucoside carbamate) led to a carbamic acid structure instead of a regioisomeric carbonate with 2-OH from the sugar moiety. Accordingly, H-1 of the glucose unit appears at 5.84 ppm and shows in the HMBC two cross peaks with C-3 of glucose at 74.0 ppm and the COOH group (158.3 ppm), each by coupling via three bonds.

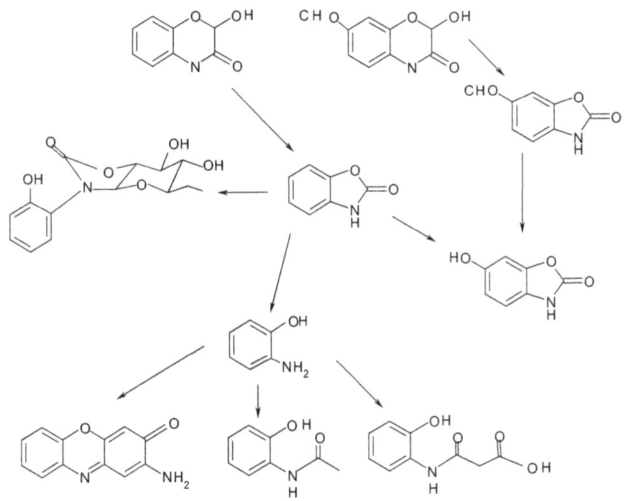

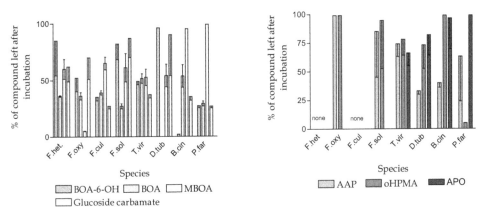

Fig. 10. Microbial degradation products derived from BOA.

Fig. 11. Mycelial plugs from agar plates (discs 0.5-1 cm in diameter) were transferred into 250 ml flasks with 100 ml sterilized Czapek medium. When mycelia were well developed, 0.5 mg were transferred to 100 ml flasks containing sterile medium without sucrose (controls) and with addition of 10 µmol BOA, BOA-6-OH, MBOA, glucoside carbamate, AAP, oHPMA, or APO (1 µmol). Cultures were grown at 25 °C in the dark without shaking. Species which did not grow without sucrose were incubated with the different compound in presence of sucrose. BOA and 2-acetamidophenol were from Aldrich, MBOA was synthesized (Sicker 1989) as well as BOA-6-OH (Wieland et al., 1999) and oHPMA (Friebe et al., 1998). Glucoside carbamate) was prepared as described (Wieland et al., 1998; Sicker et al., 2001). The cultures were harvested after 14 days of cultivation. Mycelia were separated by filtration through 100µm nylon nets, dried between paper sheets and weighted. The medium was extracted with ethyl acetate. The organic and aqueous phases were evaporated to dryness, the residues dissolved in 70 % methanol and analyzed by HPLC. $N = 5$.

The following data could be obtained by MS-analysis: In the positive ion mode (with addition of formic acid for a better ionization), several peaks appear: a protonated monomer ion at 298.09216 da (exact theoretical mass 298.09213 da) besides two sodium-adducts of appropriate mono- and dimer ions at 320.07465 da (exact theoretical mass 320.07407 da) and 617.15936 da (exact theoretical mass 617.15892 da), respectively. By addition of sodium formate to the sample solution, the two last-mentioned signals increase to the most intensive peaks in the spectrum, accompanied by a further dimer peak at 639.14197 da ([2M-H,+2Na]$^+$, exact theoretical mass 639.14087 da).

In the negative ion mode, applied to the initial methanolic solution without buffer, a corresponding weak signal at 296.07845 da (exact theoretical mass 296.07758 da) appears. With ammonia as buffer this monomer signal at 296.07806 da increases and is still accompanied by a weak dimer signal. By use of a stronger base like triethylamine, the above mentioned mono- and dimers appear again, however, now accompanied by an additional ion pair of low intensity at 314.08320 da and 629.13868 da. The latter ion was already detected in the ammonia-spectra. This at first glance odd behavior can be easily understood as follows: Object of investigation is compound **1** from a well separated peak of the HPLC chromatogram. The retention time of **1** is distinctively different from that of the precursor glucoside carbamate. Hence, our findings from the mass spectra lead to the conclusion, that under the ESI conditions the carbamic acid **1** reacts almost completely back to the glucoside carbamate by dehydration. Only under strongly basic conditions signals for the intrinsic carbamic acid with the formula $C_{13}H_{17}NO_8$ appear. Thus, by means of the MS and NMR data analysis the new compound was identified as N-ß-D-glucopyranosyl-N-(2′-hydroxyphenyl) carbamic acid (**1**), (N-glucosylated carbamic acid, Fig. 12).

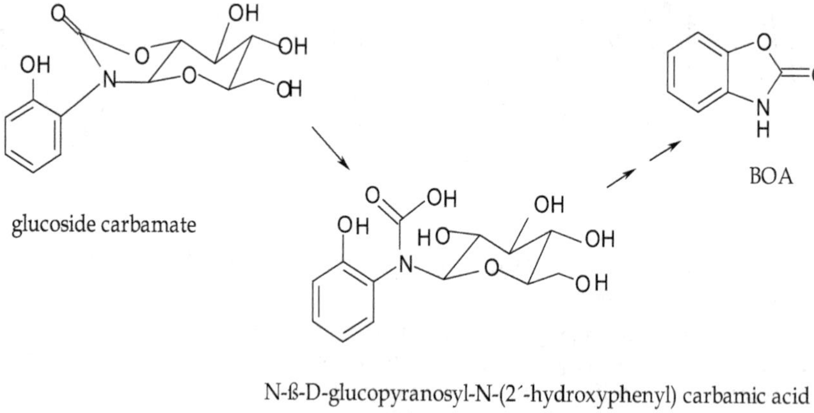

glucoside carbamate

BOA

N-ß-D-glucopyranosyl-N-(2′-hydroxyphenyl) carbamic acid

Fig. 12. Glucoside carbamate is hydrolysed to N-glucosylated carbamic acid, than deglucosylated and rearranged back to BOA. N-glucosylated carbamic acid was isolated the medium of the species mentioned in the text and purified by HPLC.

The carbamic acid feature is a rare one among natural products. Hitherto, only five representatives are known. Pallidin, a N-carboxyindole alkaloid, has been isolated from the sponge *Rhaphisia pallid* (Su et al., 1996)). Echinosulfone A, isolated from a Southern Australian marine sponge *Echinodictyum*, is a related derivative of the N-carboxyindole moiety (Ovenden et al., 1999). 1,2-Pyrrolidinedicarboxylic acid was identified as constituent of propolis balsam (Greenaway et al., 1991). N-1′-carboxybiotin has been studied in respect of the formation of enzyme N1'-carboxybiotin complexes during biochemical transformations (Jockel et al., 2000; Legge et al., 1996). However, the feature of a glycosyl carbamic acid as found in **1** has not at all been described for natural products. The most similar compound reported is synthetic ß-D-glucopyranosyl carbamic acid (Ulsperger et al., 1958). The identification of compound 1 was possible with the help of Diana Hofmann (Universität Leipzig, Institut für Analytische Chemie) and Lothar Hennig (Universität Leipzig, Institut für Organische Chemie).

8. Degradation of 2-amino-3H-phenoxazin-3-one (APO)

Fungi differ considerably in their sensibility to APO (Fig. 13). A low sensitivity is correlated with the ability to decompose the compound. All tested strains of the *Fusarium* species, *Drechslera tuberosa* and *Trichoderma viride* are able to degrade APO. With the *Fusarium* species the compound disappeared completely. The medium of *T. viride* contains traces of several phenoxazinones, indicating that some APO is modified by substitutions. *Botrytis cinerea*, which was found to be highly sensitive to APO in the growth tests, has only a low activity to degrade the compound as well as the most sensitive species *Paecilomyces farinosus*.

Since it is rather likely that APO degradation is started by oxidation, we performed experiments to elucidate how fungi can initiate oxidation processes that result in APO destruction. When 200 nmol APO was incubated with H_2O_2 in methanolic solution at room temperature, no decrease or precipitation of the compound was observed over a period of 24 h. The same result is obtained when Czapek medium (contains 10 mg $FeSO_4$/l) is used without addition of H_2O_2 (Tab. 2). Czapek medium with H_2O_2, however, led to an almost complete destruction of APO within 24 h via several intermediates. Thus, in combination with H_2O_2 and Fe^{2+} (Tab. 2), APO is easily destroyed via Fenton reaction: $Fe^{2+} + H_2O_2 \rightarrow Fe^{3+} + \cdot OH + OH^-$, in which the mechanism of radical production is still a matter of debate.

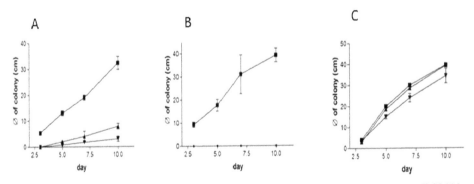

Fig. 13. Examples for the differences in the sensitivity to APO. **A**: *Botrytis cinerea* F-00646; **B**: *Paecilomyces farinosus* F-01073 and **C**: *Fusarium avenaceum* F-00475. ■: control; ▲1 µmol APO;▼2µmol APO in Czapek medium. Growing fungal mycelium was placed on the center of each Petri dish (20 cm i.d.) and incubated in the dark at 25ºC. Fungal growth was measured at the 3rd ,5th ,7th and 10th day of growth. Each experiment was repeated three times. *P. farinosus* was completely inhibited by APO, *B. cinerea* strongly inhibited by both APO concentrations. *F. avenaceum* showed no inhibition.

Hyde & Wood (1997) reported on the presence of a Fe(II) oxalate complex in aerobic solution that can lead to hydroxyl radicals without any other source. Therefore we examined the excretion of oxalate by the two fungi: *Fusarium heterosporum* F-00195 as a strain able to degrade APO rather fast and *Paecilomyces farinosus* F-01073 as a strain with a low degradation activity. Oxalate determination was done with the Trinity biotech oxalate kit No. 591.The mediums of the fungi were analyzed for secreted oxalate during the first 8 h of incubation in presence of APO, the one of *F. heterosporum* was also tested for the presence of H_2O_2 (National Diagnostics hydrogen peroxide assay kit Cl-204). During the incubation period, the pH of the media was lowered from pH 7.0 to 5.5 (*F. heterosporum*) and 5.7 (*P. farinosus*). *F. heterosporum* started to secrete oxalate already 30 min after start of the incubation (Fig. 14-16). The amount strongly increased during the next hour, but drops later on. Oxalate secretion by *P. farinosus* was measurable after 4 h of incubation, but reached only about 10% of the highest *F. heterosporum* value after 8 h. The oxalate excretion profile of *F. heterosporum* corresponds to the disappearance of APO already 1 h after starting the incubation. Contrarily, *P. farinosus* started some APO degradation after 8 h. In the *F. heterosporum* medium, H_2O_2 was measurable immediately after start of the incubation and again 8 h later. Hydrogen peroxide was extremely low during the major phase of oxalate release and APO degradation. It is assumed that oxalate excretion and the release of H_2O_2 are causative for the APO degradation of all other APO insensitive *Fusarium* strains tested.

Incubation time	APO (200 nmol)	APO + H_2O_2 (35 mmol)	APO + H_2O2 + Fe^{2+} (100 nmol)	APO + Czapek medium	APO + Czapek medium + H_2O_2	APO + 5 µmol oxalate + 5 µmol Fe^{2+}	Assay volume: 400 µl
Start (t 0)	1.02 nmol	1.15	0.96	1.20	1.24	1.20	Volume
End (t 24h)	1.04 nmol	1.16	0.30	1.25	0.16	0.70	analyzed by HPLC: 2 µl

Table 2. Destruction of APO in presence of H_2O_2 or oxalate and Fe^{2+} and controls. Average values of 5 independent experiments.

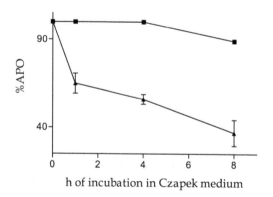

h of incubation in Czapek medium

—▲— APO-insensitive *Fusarium heterosporum* F-00195

—■— APO-sensitive *Paecilomyces farinosus* F-01073

Fig. 14. Decrease of APO in the medium of *F. heterosporum* and *P. farinosus*. N = 3.

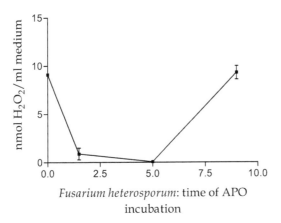

Fusarium heterosporum: time of APO
incubation

Fig. 15. During the major degradation phase, H_2O_2 is poorly present in the medium. $N = 3$.

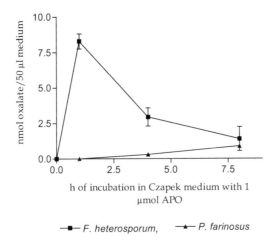

h of incubation in Czapek medium with 1
µmol APO

—■— *F. heterosporum*,   —▲— *P. farinosus*

Fig. 16. High excretion of oxalate by *F. heterosporum* during the major phase of APO degradation. $N = 3$.

Many fungi are known to secrete oxalate, some in mmolar concentrations under certain conditions (Cessna et al., 2000; Dutton & Evans, 1996). According to Varela & Tien (2003) oxalate facilitates hydroxyl radical formation at low concentration. Oxalate sequestering of ferric ions are also discussed as a protection of the fungi. Moreover, certain fungi are known to release H_2O_2. According to our study, oxalate must have a function in radical formation because oxalate can replace H_2O_2 (Tab. 2). Figure 15 illustrates that the depletion of H_2O_2 during the major phase of *F. heterosporum* APO degradation is accompanied by a strong excretion of oxalate. Thus, existing H_2O_2 is used for APO degradation and excreted oxalate lead to new hydroxyl radicals.

The oxidation of BOA, BOA-6-OH, glucoside carbamate, AAP and oHPMA was also tested (100 µmol each) for degradation via Fenton reaction. None of these compounds was destroyed in Czapek medium supplemented with H_2O_2. We assume the participation of exuded fungal enzymes that modify these compounds prior to APO production and the subsequent oxidative destruction of phenoxazinones by the Fenton reaction. It is concluded that the complete biodegradation of BOA detoxification products and phenoxazinones is a concerted action of various fungi with different metabolic properties which is probably supported by bacteria. At least the sugar moiety of the plant detoxification products can be metabolized by the microorganisms. Interestingly, Chen et al. (2010) found an increase in soil fungi after DIMBOA and MBOA application. The authors assume affects on the soil microbial community structure with a change in fungi populations in the wheat rhizosphere. Saunders and Kohn (2009) described a significant influence of benzoxazinoids on fungal endophyte communities.

Phenoxazinones are not only compounds originated from benzoxazinone degradation. They are synthesized by a variety of different organisms, for instance by members of the genus *Pycnoporus* (Sullivan & Henry 1971; Temp & Eggert 1998) or by *Streptomyces* species (Suzuki et al., 2006). Clearly, these compounds are relatively wide-spread in nature. Macias et al. (2005) reported on a life time of APO of more than 80 days in some soils. Unfortunately, neither the source of the soil nor its quality is mentioned. Also Bacon et al. (2007) take APO for a stable compound. In contrast, Krogh et al. (2006) determined 1.4×10^{-11} M APO as the highest concentration in sandy loam soil after incorporation of one rye seedling 300 mg[1]- soil. The APO concentration decreased rapidly during 10 days to about 30%.

For APO degradation in nature, several soil properties such as a high diversity of microorganisms including ones that excrete chelator agents (e.g. oxalate), generation of H_2O_2, the presence of iron, and a pH lower than 6 are certainly prerequisites for starting APO degradation via Fenton reaction. Even if the life time is variable depending on the soil conditions, APO is fortunately not a stable compound but an allelochemical which can be completely degraded over time. The Fenton reaction is the key reaction in the oxidation of membrane lipids, oxidation of amino acids and biologic reactions where biological reduction agents are present. The reaction is common in chemical, biological, and environmental systems (Barbusinski, 2009; Prousek, 2007). The importance of the Fenton reaction in natural environments and in waste treatments for degradation of phenolic and other compounds has been recognized during the last two decades (Pignatello et al., 2006; Vlyssides et al., 2011).

9. Conclusion

The bio-accumulation of conventional herbicides/pesticides and their often highly toxic degradation products is a well-known problem. Another problem is their persistence in soil and ground water. Some of these compounds are not only genotoxic, carcinogenic, neurotoxic and immunotoxic but have also negative effects on the fertility of vertebrates and are toxic to bees. Moreover, plants have developed new strategies in the resistance against common herbicides (Gainesa et al., 2010; Powles &Yu, 2010). Yuan et al. (2006) summarize in their article the dramatic increase in herbicide-resistant weed biotypes which became obvious since the late 80ies of the last century. This demands innovative and environmental-friendly strategies based on sustainable resources using natural, plant–own compounds for weed control which are acceptable by consumers. Breeding crops suitable for natural product applications has to be aimed. Allelopathic concepts are more and more attractive in agriculture. However a number of prerequisites are of importance. Among others, the selectivity and the relatively fast and complete biodegradation of the compounds is of outstanding importance to avoid environmental damage and the destruction of biodiversity. The same priority has the development of new agricultural systems. Applications of large quantities of a natural compound, perhaps booted by artificial substitutions, instead of conventional herbicides, cannot be a solution. The same problems as obtained with common herbicides will occur soon. Therefore, the reactions of plant, including detoxification strategies, of microorganisms, of plant and microbial genotypes, soil properties, fertilizer management, occurrence and degradation of the mother substances and their derivatives have to be investigated carefully prior to the use of any natural compound as a bioherbicide. This knowledge will help to design cultural systems promoting natural compounds to agricultural crops with a beneficial impact on environmental safety as well as consumers' health. Benzoxazinoids could be a group of allelochemicals that seem to fulfill several of the prerequisites. The "journey" of benzoxazolinone and its final degradation demonstrates at least some of the advantages of the use of suitable natural bioactive compounds in agriculture: highly selective efficiency and a short existence (Fig. 17). It gives also an impression of the complexity of the ecological net involved in the transformation and degradation of an allelochemical, although only a few aspects could be directed in this article.

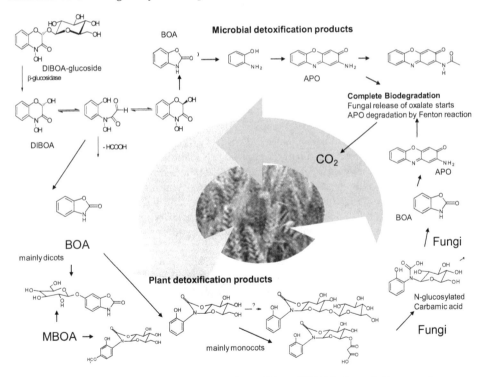

Fig. 17. Illustration of the journey. BOA is released from DIBOA degradation, can be absorbed by other plants, for example, weeds or individuals of the same species (rye, wheat, maize). Plant and microbial detoxification products can be exuded and are converted by defined microorganisms. Phenoxazinone(s) can be completely degraded via Fenton reaction.

10. References

Ahmad, S., Veyrat, N., Gordon-Weeks, R., Zhang, Y., Martin, J., Smart, L., Glauser, G., Erb, M., Flors, V., Frey, M.& Ton, J. (2011). Benzoxazinoid metabolites regulate innate immunity against aphids and fungi in maize. *Plant Physiol. 2011 pp.111.180224*

Anai, T., Aizawa, H., Ohtake, N., Kosemura, S., Yamamura, S. & Hasegawa, K. (1996). A new auxin inhibiting substance, 4-Cl-6,7-dimethoxy-2-benzoxazolinone, from light-grown maize shoots. *Phytochemistry* 42, 273–275.

Bacon, C.W., Hinton, D.M., Glenn, A.E., Macias, F.A. & Marin, D. (2007). Interactions of *Bacillus mojavensis* and *Fusarium verticillioides* with a benzoxazolinone (BOA) and its transformation product, APO. *J.Chem. Ecol.* 33, 1885-1897.

Baerson, S.R., Sanchez-Moreiras, A.M., Pedrol-Bonjoch, N., Schulz, M., Kagan, I.A., Agarwal, A.K., Reigosa, M.J. & Duke, S.O. (2005). Detoxification and transcriptome response in *Arabidopsis* seedlings exposed to the allelochemical benzoxazolin-2(3*H*)-one. *J. Biol. Chem.* 280, 21867-21881.

Baluška, F., Vitha, S., Barlow, P.,W. & Volkmann, D. (1997). Rearrangements of F-actin arrays in growing cells of intact maize root apex tissues: a major developmental switch occurs in the postmitotic transition region. *Eur. J. Cell Biol.* 72, 113-121.

Baluška, F., Busti, E., Dolfini, S., Gavazzi, G. & Volkmann, D. (2001b). *Lilliputian* mutant of maize shows defects in organization of actin cytoskeleton. *Dev. Biol.* 236, 478-491.

Baluška, F., Barlow, P.,W. & Volkmann, D. (2000). Actin and myosin VIII in developing root cells. In: *Actin: a Dynamic Framework for Multiple Plant Cell Functions*. Staiger CJ, Baluška F, Volkmann D, Barlow PW (eds), pp. 457-476, Kluwer Academic Publishers, Dordrecht, The Netherlands,.

Baluška, F., Jásik, J., Edelmann, H.G., Salajová, T. & Volkmann, D. (2001a). Latrunculin B induced plant dwarfism: plant cell elongation is F-actin dependent. *Dev. Biol.* 231, 113-124.

Baluška, F., Mancuso, S., Volkmann, D. & Barlow, P.W. (2010). Root apex transition zone: a signalling – response nexus in the root. *Trends Plant Sci.* 15, 402-408.

Baluška, F., Vitha, S., Barlow, P.W. & Volkmann, D. (1997). Rearrangements of F-actin arrays in growing cells of intact maize root apex tissues: a major developmental switch occurs in the postmitotic transition region. *Eur. J. Cell Biol.* 72, 113-121.

Baluška, F., Volkmann, D. & Barlow, P.W. (2001d). A polarity crossroad in the transition growth zone of maize root apices: cytoskeletal and developmental implications. *J. Plant Growth Regul.* 20, 170-181.

Baluška, F., Volkmann, D. & Menzel, D. (2005). Plant synapses: actin-based adhesion domains for cell-to-cell communication. *Trends Plant Sci.* 10, 106-111.

Baluška, F., Šamaj, J., Wojtaszek, P., Volkmann, D. & Menzel, D. (2003b). Cytoskeleton – plasma membrane – cell wall continuum in plants: emerging links revisited. *Plant Physiol.* 133, 482-49.

Baluška, F. & Hlavacka, A. (2005). Plant formins come to age: something special about cross-walls. *New Phytol.* 168, 499-503.

Baluška, F., von Witsch, M., Peters, M., Hlavačka, A. & Volkmann, D. (2001c). Mastoparan alters subcellular distribution of profilin and remodels F-actin cytoskeleton in cells of maize root apices. *Plant Cell Physiol.* 42, 912-922.

Baumeler, A., Hesse, M. & Werner, C. (2000). Benzoxazinoids-cyclic hydroxamic acids, lactams, and their corresponding glucosides in the genus *Aphelandra* (Acanthaceae). *Phytochemistry* 53, 213-222.

Barbusinski, K. (2009). Fenton reaction – controversy concerning the chemistry. *Ecol. Chem. Engin.* 16, 347-358.

Barnes, J.P. & Putnam, A.R. (1987). Role of benzoxazinones in allelopathy by rye (*Secale cereale* L.). *J. Chem. Ecol.* 56, 1788-1800.

Batish, D.R., Singh, H.P., Setia, N., Kaur, S. & Kohli, R.K. (2006). 2-Benzoxazolinone (BOA) induced oxidative stress, lipid peroxidation and changes in some antioxidant enzyme activities in mung bean (*Phaseolus aureus*). *Plant Physiol. Biochem.* 44, 819-827.

Burdziak, A., Schulz, M. & Klemme, J.H. (2001). Isolation and characterization of bacteria being able to degrade benzoxazolinone (BOA). In: *First European Allelopathy Symposium-Physiological Aspects of Allelopathy*. Reigosa, M.J., Bonjoch, N.,P. (eds.) p. 224, Gamesal, ISBN 84-95046-18-0, Vigo, Spain.

Burgos, N.R., Talbert, R.E. & Mattice, J.D. (1999). Cultivar and age differences in the production of allelochemicals by *Secale cereale*. *Weed Sci.* 47, 481-485.

Burgos, N.R., Talbert R.E., Kim, K.S. & Kuk, Y.I. (2004). Growth inhibition and root ultrastructure of cucumber seedlings exposed to allelochemicals from rye (*Secale cereale*). *J. Chem. Ecol.* 30, 671-689.

Carter, J.P., Spink, J., Cannon, P.F., Danils, M.J., Osbourn, A,E. (1999). Isolation, characterization, and avenacin sensitivity of a diverse collection of cereal-root-colonizing fungi. *Appl. Environ. Microbiol.* 65, 3364-3372.

Cessna, S.G., Sears, V.E., Dickman, M.B. & Low, P.B. (2000). Oxalic acid, a pathogenicity factor for *Sclerotinia sclerotiorum* , suppresses the oxidative burst of the host plant. *The Plant Cell* 12, 2191–2199.

Chen, K.-J., Zheng, Y.-Q., Kong, C.-H., Zhang, S.-Z., Li, J. & Liu, X.-G. (2010). 2,4-Dihydrox-7-methoxy-1,4-benzoxazin-3-one (DIMBOA) and 6-methoxy-benzoxalin-2-one (MBOA) levels in the wheat rhizosphere and their effect on the soil microbial community structure. *J. Agric. Food Chem.* 58, 12710-12716.

Dhonukshe, P., Grigoriev, I., Fischer, R., Tominaga, M., Robinson, D.G., Hasek, J., Paciorek, T., Petrásek, J., Seifertová, D., Tejos, R., Meisel, L.A., Zazímalová, E., Gadella, T.W., Jr., Stierhof, Y.D., Ueda, T., Oiwa, K., Akhmanova, A., Brock, R., Spang, A. & Friml, J. (2008). Auxin transport inhibitors impair vesicle motility and actin cytoskeleton dynamics in diverse eukaryotes. *Proc Natl Acad Sci USA* 105, 4489-4494.

Dyachok, J., Shao, M.R., Vaughn, K., Bowling, A., Facette, M., Djakovic, S., Clark, L. & Smith, L. (2008). Plasma membrane-associated SCAR complex subunits promote cortical F-actin accumulation and normal growth characteristics in *Arabidopsis* roots. *Mol. Plant* 1, 990-1006.

Dixon, D.P., Skipsey, M. & Edward, R. (2010). Roles for glutathione transferases in plant secondary metabolism. *Phytochemistry* 71, 338-350.

Dutton, M.V. &. Evan, C.S. (1996) Oxalate production by fungi: its role in pathogenicity and ecology in the soil environment. *Can. J. Microbiol.* 42, 881-895.

Frébortová, J., Novák, O., Frébort, I. & Jorda, R. (2010) Degradation of cytokinins by maize cytokinin dehydrogenase is mediated by free radicals generated by enzymatic oxidation of natural benzoxazinones. *The Plant J.* 61, 467-481.

Frey, M., Chomet, P., Glawischnig, E., Stettner, C., Grün, S., Winklmair, A., Eisenreich, W., Bacher, A., Meerley, R.B., Briggs, S.P., Simcox, K. & Gierl, A. (1997). Analysis of a chemical plant defense mechanism in grasses. *Science* 277, 696-699.

Frey, M., Schullehner, K., Dick, R., Fiesselmann, A. & Gierl, A. (2009). Benzoxazinoid biosynthesis, a model for evolution of secondary metabolic pathways in plants. *Phytochemistry* 70, 1645-1651.

Friebe, A., Roth, U., Kück, P., Schnabl, H. & Schulz, M. (1997). Effects of 2,4-dihydroxy-1,4-benzoxazin-3-ones on the activity of plasma membrane H+-ATPase. *Phytochemistry* 44, 979-983.

Friebe, A., Vilich, V., Hennig, L., Kluge, M. & Sicker D. (1998). Detoxification of benzoxazolinone allelochemicals from wheat by *Gaeumannomyces graminis* var. *tritici, G. graminis* var. *graminis, G. graminis* var. *avenae*, and *Fusarium culmorum*. *Appl. Environ. Microbiol.* 64, 2386-2391.

Gaines, T., A., Zhang, W., Wang, D., Bukun, B., Chisholm, S., T., Shaner, D., L., Nissen, S., J., Patzoldt, W., L., Tranel, P., J., Culpepper, A., S., Grey, T., L., Webster, T., M., Vencill, W., K., Sammons, R., D., Jiang, J., Preston, C., Leach, J., E. & Westra, P. (2010). Gene amplification confers glyphosate resistance in *Amaranthus palmeri*. PNAS 107,1029-1034.

Gamczarka, M. (2005). Response of the ascorbate-glutathione cycle to re-aeration following hypoxia in lupine roots. *Plant Physiol. Biochem.* 43, 583-590.

Gavazzi, C., Schulz, M., Marocco, A. & Tabaglio, V. (2010). Sustainable weed control by allelochemicals from rye cover crops from greenhouse to field evidence. *Allelopathy J.* 25, 259-274.

Glawischnig, E., Grün, S., Frey, M. & Gierl, A. (1999). Cytochrome P450 monooxygenase of DIBOA biosynthesis: specificity and conservation among grasses. *Phytochemistry* 50, 925-930.

Glenn, A.E., Gold, S.E. & Bacon, C.W. (2002). *Fdb1* and *Fdb2, Fusarium verticillioides* loci necessary for detoxification of preformed antimicrobials from corn. *Molec. Plant-Microbe Interact.* 15, 91-101.

Glenn, A.E., Hinton, D.M., Yates, I.E. & Bacon, C.W. (2001). Detoxification of corn antimicrobial compounds as the basis for isolating *Fusarium verticillioides* and some other *Fusarium* species from corn. *Appl. Environ. Microbiol.* 67, 2973-2981.

Glenn, A.S., Meredith, F.I., Morrison III, W.H. & Bacon, C.W. (2003). Identification of intermediate and branch metabolites resulting from biotransformation of 2-benzoxazolinone by *Fusarium verticillioides*. *Appl. Environ. Microbiol.* 69, 3165-3169.

Gianoli, E. & Niemeyer H.M. (1997). Environmental effects on the accumulation of hydroxamic acids in wheat seedlings: The importance of plant growth rate. *J.Chem. Ecol.* 23, 543-551.

Greenaway, W., May, J., Scaysbrook, T. & Whatley, F.R. (1991). Identification by gas chromatography-mass spectrometry of 150 compounds in propolis. *Zeitsch. Naturforsch. C J. Biosciences* 46,111-121.

Habig, W.H. & Jacoby, W.B. (1981).Assays for differentiation of glutathione-S-transferases. *Method. Enzymol.* 77, 398-405.

Hochholdinger, F., Zimmermann, R. (2008) Conserved and diverse mechanisms in root development. *Curr. Opin. Plant Biol.* 11, 70-74.

Hofman, N.D., Knop, M., Hao, H., Hennig, L., Sicker, D. & Schulz, M. (2006). Glucosides from MBOA and BOA detoxification by *Zea mays* and *Portulaca oleracea*. *J. Nat. Prod.* 69, 34-37.

Hohenberger, P., Eing, C., Straessner, R., Durst, S., Frey, W. & Nick, P. (2011). Plant actin controls membrane permeability. *Biochim. Biophys. Acta* 1808, 2304-2312.

Hoshi-Sakoda, M., Usu, K., Ishizuka, K., Kosemura, S., Yamamura, S. & Hasegawa, K. (1994). Structure-activity relationships of benzoxazolinones with respect to auxin-induced growth and auxin-binding protein. *Phytochemistry* 37, 297-300.

Hruz, T., Laule, O., Szabo, G., Wessendorp, F., Bleuer, S., Oertle, L., Widmayer, P., Gruissem, W. & Zimmermann, P. (2008). Genevestigator V3: a reference expression database for the meta-analysis of transcriptomes. *Adv. Bioinformat.*, article ID 420747,5 pages. Doi:10.1155/2008/420747.

Hyde, S.M. & Wood, P.M. (1997). A mechanism for production of hydroxyl radicals by the brown-rot fungus *Coniophora puteana*: Fe(III) reduction by cellobiose dehydrogenase and Fe(II) oxidation at a distance from the hyphae. *Microbiol.* 143, 259-266.

Jockel, P., Schmid, M., Choinowski, T. & Dimroth, P. (2000). Essential role of tyrosine 229 of the oxaloacetate decarboxylase beta-subunit in the energy coupling mechanism of the Na$^{(+)}$ pump. *Biochemistry* 39, 4320-4326.

Jonczyk R., Schmidt, H., Osterrieder, A., Fisselmann, A., Schullehner, K., Haslbeck, M., Sicker, D., Hofmann, D., Yalpani, N., Simmons, C. Frey, M. & Gierl, A. (2008). Elucidation of the final reactions of DIMBOA-glucoside biosynthesis in maize: characterization of *Bx6* and *Bx7*. *Plant Physiol.* 146, 1053-1063.

Knop, M., Pacyna, S., Voloshchuk, N., Kant, S., Müllenborn, C., Steiner, U., Kirchmair, M., Scherer, H.W. & Schulz, M. (2007). *Zea mays* : benzoxazolinone detoxification under sulfur deficiency conditions - a complex allelopathic alliance including endophytic *Fusarium verticillioides*. *J. Chem. Ecol.* 33, 225-237.

Kocsy, G., von Ballmoos, S., Suter, M., Ruegsegger, A., Galli, U., Szalai, G., Galiba, G. & Brundhold, C. (2000). Inhibition of glutathione synthesis reduces chilling tolerance in maize. *Planta* 211, 528-536.

Krogh, S.S., Mensz, S.J.M., Nielsen, S.T., Mortensen, A.G., Christophersen, C. & Fomsgaard, I.S. (2006). Fate of benzoxazinone allelochemicals in soil after incorporation of wheat and rye sprouts. *J. Agric. Food. Chem.* 54, 1064-1074.

Legge, G.B., Branson, J.P. & Attwood, P.V. (1996). Effects of acetyl CoA on the pre-steady-state kinetics of the biotin carboxylation reaction of pyruvate carboxylase. *Biochemistry* 35, 3849-3856.

Macias, F.A., Oliveros-Bastidas, A., Marin, D., Castellano, D., Simonet , A.M. & Molinillo J.M.G. (2005). Degradation studies on benzoxazinoids. Soil degradation dynamics of (2*R*)-2-ß-D-glucopyranosyl-4-hydroxy(2*H*)-1,4-phenoxazin-3-(4*H*)-one (DIBOA-Glc) and its degradation products, phytotoxic allelochemicals from Gramineae. *J. Agricult. Food Chem.* 53, 538-548.

Mancuso, S., Marras, A.M., Volker, M. & Baluška, F. (2005). Non-invasive and continuous recordings of auxin fluxes in intact root apex with a carbon-nanotube-modified and self-referencing microelectrode. *Anal. Biochem.* 341, 344-351.

Mancuso, S., Marras, A.M., Mugnai, S., Schlicht, M., Zarsky, V., Li, G., Song, L., Hue, H.W. & Baluška, F. (2007). Phospholipase Dζ2 drives vesicular secretion of auxin for its polar cell-cell transport in the transition zone of the root apex. *Plant Signal Behav.* 2, 240-244.

Niemeyer, H.M. (2009). Hydroxamic acids derived from 2-hydroxy-2*H*-1,4-benzoxazin-3(4*H*)-one: Key defense chemicals ofcereals. *J. Agric. Food Chem.* 57, 1677-1696.

Noctor, G., Queval, G., Mhamdi, A., Charouch, S. & Foyer C.H. (2011). Glutathione. The *Arabidopsis* Book, 9, 1-32.

Ovenden, S.P.B. & Capon, R.J. (1999). Echinosulfonic acids A-C and Echinosulfone A: novel bromoindole sulfonic acids and a sulfone from a Southern Australian marine sponge, *Echinodictyum. J. Nat. Prod.* 62, 1246-1249.

Péret, B., De Rybel, B., Casimiro, I., Benková, E., Swarup, R., Laplaze, L., Beeckman, T. & Bennett, M.J. (2009). *Arabidopsis* lateral root development: an emerging story. *Trends Plant Sci* 14, 399-408.

Pignatello J.J., Oliveros, E. & MacKay, A. (2006). Advanced oxidation processes for organic contaminant destruction based on the Fenton reaction and related chemistry. *Critic. Rev. Environ. Sci. & Technol.* 36, 1-84.

Powles,S., B. & Yu, Q. (2010). Evolution in action: Plants resistant to herbicides. Annu. Rev. Plant Biol. 61, 317-347.

Prousek, J. (2007). Fenton chemistry in biology and medicine. *Pure Appl. Chem.*, 79, 2325-2338.

Rahman, A., Bannigan, A., Sulaman, W., Pechter, P., Blancaflor, E.B & Baskin, T.I. (2007). Auxin, actin and growth of the *Arabidopsis thaliana* primary root. *Plant J* 50, 514-528.

Reberg-Horton, S.C., Burton, J.D., Danehower, D.A., Ma, G., Monks, D.W., Murphy, J.P., Ranells, N.N., Williamson, D.J. & Creamer, N.G. (2005). Changes over time in the allelochemical content of ten cultivars of rye (*Secale cereale* L.). *J. Chem. Ecol.* 31, 179-193.

Rice, C.P., Park, Y.B., Adam, F., Abdul-Baki, A.A. & Teasdale, J.R. (2005). Hydroxamic acid content and toxicityof rye at selected growth stages. *J. Chem. Ecol.* 31, 1887-1905.

Richardson, M.D. & Bacon, C.W. (1993). Cyclic hydroxamic acid accumulation in corn seedlings exposed to reduced water potentials before, during and after germination. *J. Chem. Ecol.* 19, 1613-1624.

Riechers, D.E., Kreuz, K., & Zhang, Q. (2010). Detoxification without Intoxication: Herbicide safeners activate plant defense gene expression. *Plant Physiol.* 153, 3-13.

Sanchez-Moreiras, A.M., Oliveros-Bastidas, A. & Reigosa, M.J. (2010). Reduced photosynthetic activity is directly correlated with 2-(3*H*)-benzoxazolinone accumulation in lettuce leaves. *J. Chem. Ecol.* 36, 205-209.

Sanchez-Moreiras, A.M., Martinez-Peñalver, A. & Reigosa, M.J. (2011). Early senescence induced by 2-3*H*-benzoxazolinone (BOA) in *Arabidopsis thaliana. J. Plant Physiol.* 168, 863-870.

Saunders, M. & Kohn, L. M. 2009. Evidence for alteration of fungal endophyte communities assembly by lost defense compounds. *New Phytol.* 182, 229-238.

Singh, H.P., Batish, D.R., Kaur, S., Setia, N. & Kohli, R.K. (2005). Effects of 2-benzoxazolinone on the germination, early growth and morphogenetic response of mung bean (*Phaseolus aureus*). Ann. Appl. Biol. 147, 267-274.

Scherer, H.W. (2001). Sulfur in crop production. *Eur. J. Agron.* 14, 81-111.

Scherer, H.W. (2009) Sulfur in soils. *J. Plant Nutr. Soil Sci.* 172, 326-335.

Schlicht, M., Strnad, M., Scanlon, M.J., Mancuso, S., Hochholdinger, F., Palme, K., Volkmann, D., Menzel, D. & Baluška, F. (2006). Auxin immunolocalization implicates vesicular neurotransmitter-like mode of polar auxin transport in root apices. *Plant Signal Behav.* 1, 122-133.

Schullehner, K., Dick, R., Vitzthum, F., Schwab, W., Brandt, W., Frey, F. & Gierl, A. (2008). Benzoxazinoid biosynthesis in dicot plants. *Phytochemistry* 69, 2668-2677.

Schulz, M. & Wieland, I. (1999). Variations in metabolism of BOA among species in various field communities – biochemical evidence for co-evolutionary processes in plant communities? *Chemoecology* 9, 133-141.

Schulz, M., Knop, M., Kant, S., Sicker, D., Voloshchuk, N. & Gryganski, A. (2006) Detoxification of allelochemicals - the case of benzoxazolin-2(3H)-one (BOA). In: *Allelopathy: a physiological Process with ecological implications.* (Reigosa, M.J., Pedrol, N. & Gonzales, L. eds), pp. 157-170, Springer, ISBN-10 1-4020-4279-5, Dordrecht, The Netherlands.

Schulz, M., Kant, S., Knop, M., Sicker, D., Colby, T., Harzen, A. & Schmidt, J. (2008). The allelochemical Benzoxazolinone – Molecular backgrounds of its detoxification and degradation. 5th World Congress on Allelopathy, p.63, Saratoga Springs, NY, USA, 21-25. September 2008.

Schulz, M., Marocco, A. & Tabaglio, V. (2011) BOA detoxification during germination and seedlings growth of for summer weeds. Submitted.

Sicker, D. (1989). A facile synthesis of 6-methoxy-2-oxo-2,3-dihydrobenzoxazole. *Synthesis* 875-876.

Sicker, D. Frey, M., Schulz, M. &, Gierl, A. (2000). Role of natural benzoxazinones in the survival strategies of plants. In: *International Review of Cytology - A Survey of Cell Biology Vol. 198.* Jeong, K.W. (ed), pp. 319-346, Academic Press, ISBN 0-12-364602-2, London.

Sicker, D., Schneider, B., Hennig, L., Knop, M. & Schulz, M. (2001). Glucoside carbamate from benzoxazolin-2(3H)-one detoxification in extracts and exudates of corn roots. *Phytochemistry* 58, 819-825.

Sicker, D. & Schulz, M. (2002). Benzoxazinones in plants: Occurrence, synthetic access, and biological activity. *Studies in Natural Products Chemistry* 27, 185-232.

Sicker, D., Hao, H., Schulz, M. (2004) Benzoxazolin-2-(3H)-ones. Generation, Effects and detoxification in the competition among plants, In: *Allelopathy-Chemistry and Mode of Action of Allelochemicals.* Macias, F.A., Galindo, J.C.G., Molinillo, J.M.G., Cutler, H.G., (eds.), pp. 77-102, CRC Press, ISBN 0-8493-1964-1, Boca Raton, Florida, USA.

Su, J., Zhong, Y., Zeng, L., Wu, H., Shen, X. & Ma, K. (1996). A new N-carboxyindole alkaloid from the marine sponge *Rhaphisia pallida. J. Nat. Prod.* 59, 504-506.

Sukuzi, H., Furusho, Y., Higashi, T., Ohnishi, Y. & Horinouch, S. (2006). A novel o-aminophenol oxidase responsible for formation of the phenoxazinone chromophore of grixazone. *J. Biol. Chem.* 281, 824-833.

Sullivan, G. & Henry, E.,D. (1971). Occurrence and distribution of phenoxazinone pigments in the genus *Pycnoporus. J. Pharm. Sci.* 60, 1097-1098.

Tabaglio, V. & Gavazzi, C. (2006). Effetti dei residui di segale sulle infestanti estive del mais. *Inf. Agr.* 62, 37-40.

Tabaglio, V., Gavazzi, C., Schulz, M. & Marocco, A. (2008). Alternative weed control using the allelopathic effect of natural benzazinoids from rye mulch. *Agron. Sust. Devel.* 28, 397-401.

Teare, J.P., Punchard, N.A., Powell, J.J., Lumb, P.J., Mitchell, W.D. & Thompson, R.P.H. (1993). Automated spectrophotometric method for determining oxidized and reduced glutathione in liver. *Clin. Chem.* 39, 686-689.

Tremp, U. & Eggert, C. (1998.: Novel interactions between laccase and cellobiose dehydrogenase during pigment *synthesis in the white rot fungus Pycnoporus cinnabarinus*. *Microbiol. 65, 389-395*.

Ulsperger, E., Bock, M. & Gradel, A. (1958). Synthesis of nonionic surface-active compounds derived from carbohydrates. *Fette, Seifen, Anstrichmittel 6,* 819-826.

Varela, E. & Tien, M. (2003). Effect of pH and oxalate on hydroquinone-derived hydroxyl radical formation during brown rot wood degradation. *Appl. Environ. Microbio.* 69, 6025-6031.

Vlyssides, A., Barampouti, E., M., Mai, S., Sotiria, M. & Eleni, N. (2011). Degradation and mineralization of gallic acid using Fenton's Reagents. *Environ. Engin. Sci.* 28, 515-520.

Von Rad, U., Hüttl, R., Lottspeich, F., Gierl, A. & Frey, M. (2001). Two glucosyltransferases involved in detoxification of benzoxazinoids in maize. *The Plant J.* 28, 633-642.

Wieland, I., Kluge, M., Schneider, B., Schmidt, J., Sicker, D. & Schulz, M. (1998). 3-ß-D-Glucopyranosyl-benzoxazolin-2(3H)-one - A detoxification product of benzoxazolin-2(3H)-one in oat roots. *Phytochemistry* 48, 719-722.

Wong, B.S., Liu, T., Ri, R., Pan, T., Petersen, R.P., Smith, M., A., Gambetti, P., Perry, G., Manson, J.C., Brown, D.R. & Sy, M.S. (2001). Increased levels of oxidative stress markers detected in the brains of mice devoid of prion protein. *J. Neurochem.* 76, 565-572.

Yuan, J., S., Tranel, P., J. & Stewart Jr, C.N. (2006). Non-target-site herbicide resistance: a family business. Trends Plant Sci., 12, 6-13.

Zasada, I.A., Rice, C.P. & Meyer, S.L.F. (2007). Improving the use of rye (*Secale cereale*) for nematode management: potential to select cultivars based on *Meloidogyne incognita* host status and benzoxazinoid content. *Nematology* 9, 53-60.

Zikmundova, M., Drandarov, K., Bigler, L., Hesse, M. & Werner, C. (2002a). Biotransformation of 2 benzoxazolinone and 2-hydroxy-1,4-benzoxazin-3-one by endophytic fungi isolated from *Aphelandra tetragona*. *Appl. Environ. Microbiol.* 68, 4863-4870.

Zikmundova, M., Drandarov, K., Hesse, M. & Werner C. (2002b). Hydroxylated 2-amino-3H-phenoxazin-3-one derivatives as products of 2-hydroxy-1,4-benzoxazin-3-one (HBOA) biotransformation by *Chaetosphaeria sp.*, an endophytic fungus from *Aphelandra tetragona*. *Zeitschrift für Naturforschung, 57C*, 660-665.

2

Fate and Determination
of Triazine Herbicides in Soil

Helena Prosen
*University of Ljubljana, Faculty of Chemistry
and Chemical Technology, Ljubljana
Slovenia*

1. Introduction

Triazine herbicides belong to the group of the most widely used herbicides worldwide. In this review paper, encompassing mostly the relevant research and publications done in the last decade, the fate of triazine herbicides after their introduction in the environment will be discussed. They are transformed in a variety of transformation products after their application, and some of these products are at least as important for the assessment of the overall fate of triazine herbicides and their impact on the environment. Both parent compounds and transformation products will be discussed with particular emphasis on their behaviour in the soil. Analytical methods for the determination of their residues and transformation products in the soil will be reviewed along with the consideration of the impact of the current analytical approaches on our knowledge about the fate of triazines.

2. Physico-chemical properties of triazines

Chemically, triazine herbicides are comprised of asymmetrical triazines (triazinones, triazidinones) and symmetrical or 1,3,5-triazines (s-triazines): chlorotriazines, methoxytriazines, methylthiotriazines. Structures of the more important triazines and their transformation products (TPs) are shown in Fig. 1.

Physico-chemical properties of compounds relevant for their behaviour in the environment are their polarity (expressed as *n*-octanol-water partitioning coefficient K_{ow}), linked to water solubility, moreover their acido-basic properties (expressed as dissociation constant K_a) and volatility (usually expressed as vapour pressure). These are listed in Table 1 for the more environmentally important triazines.

3. Toxicity and environmental effects of triazines

Triazine herbicides are generally of low acute toxicity for birds and mammals, although certain species show unexpected vulnerability for some of them, e.g. for sheep the fatal dose of simazine has been reported as 500-1400 mg/kg, while LD_{50} for rats is >5000 mg/kg (Stevens & Sumner, 1991). Acute toxicity data for some compounds are shown in Table 2.

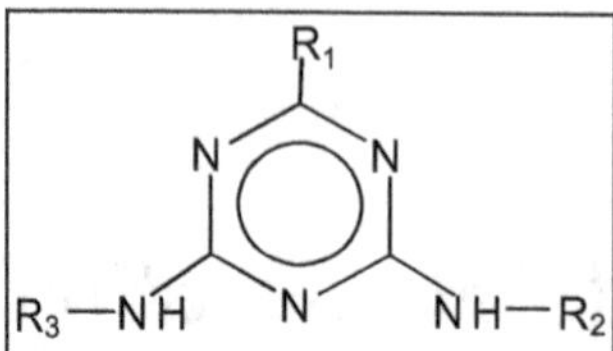

		R_1	R_2	R_3
General s-triazine structure (see table on the right)	Chlorotriazines			
	atrazine	Cl	CH_2CH_3	$CH(CH_3)_2$
	simazine	Cl	CH_2CH_3	CH_2CH_3
	propazine	Cl	$CH(CH_3)_2$	$CH(CH_3)_2$
	cyanazine	Cl	CH_2CH_3	$C(CH_3)_2CN$
	terbutylazine	Cl	CH_2CH_3	$C(CH_3)_3$
	Methoxytriazines			
	atratone	OCH_3	CH_2CH_3	$CH(CH_3)_2$
	prometon	OCH_3	$CH(CH_3)_2$	$CH(CH_3)_2$
	terbumeton	OCH_3	CH_2CH_3	$C(CH_3)_3$
metribuzin (triazinone)	Methylthiotriazines			
	ametryn	SCH_3	CH_2CH_3	$CH(CH_3)_2$
	simetryn	SCH_3	CH_2CH_3	CH_2CH_3
	prometryn	SCH_3	$CH(CH_3)_2$	$CH(CH_3)_2$
	terbutryn	SCH_3	CH_2CH_3	$C(CH_3)_3$
	Degradation products			
hexazinone (triazidinone)	desethylatrazine	Cl	H	$CH(CH_3)_2$
	desisopropylatrazine	Cl	CH_2CH_3	H
	desethyldesisopropylatrazine	Cl	H	H
	hydroxyatrazine	OH	CH_2CH_3	$CH(CH_3)_2$

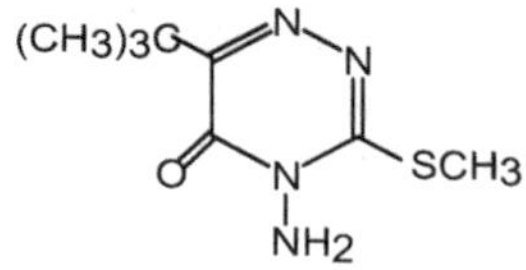

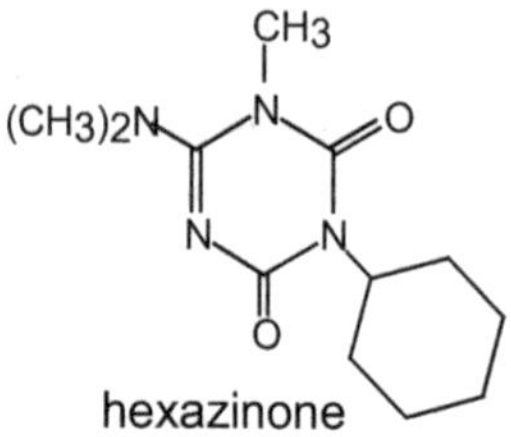

Fig. 1. Structures of some widely used triazines and more important transformation products.

Name	M / g/mol	Water sol. / mg/L	$\log K_{ow}$	pK_a	p / Pa
Atrazine	215.7	33 (20 °C)	2.2-2.7	1.7	$4.0 \cdot 10^{-5}$ (20 °C)
Simazine	201.7	5 (20-22 °C)	2.2-2.3	1.65	$8.1 \cdot 10^{-7}$ (20 °C)
Cyanazine	240.7	171 (25 °C)	1.8-2.0	1.85	$2.1 \cdot 10^{-7}$ (25 °C)
Terbutylazine	229.8	8.5 (20 °C)	2.6-3.0	2.0	$1.5 \cdot 10^{-4}$ (25 °C)
Atraton	211.3	1800 (20-22 °C)	2.3-2.7	4.2	NA
Terbumeton	225.3	130 (20 °C)	2.7-3.1	4.7	$2.5 \cdot 10^{-5}$ (25 °C)
Ametryn	227.1	185 (20 °C)	2.7-3.1	4.0	$1.1 \cdot 10^{-4}$ (20 °C)
Prometryn	241.4	33-48 (20 °C)	3.3	4.1	$1.3 \square 10^{-4}$ (20 °C)
Terbutryn	241.4	25 (20 °C)	3.1-3.7	4.3	$2.2 \cdot 10^{-4}$ (25 °C)
Desethylatrazine	187.7	3200	1.5	1.65	$1.2 \cdot 10^{-2}$ (25 °C)
Desisopropylatrazine	173.6	670	1.1-1.2	1.58	NA
Hydroxyatrazine	197.3	5.9	1.4	5.2	$1.1 \cdot 10^{-3}$ (25 °C)

Table 1. Some relevant physico-chemical parameters for the environmentally important triazines and their transformation products (Kaune et al., 1998; Noble, 1993; Shiu et al., 1990; Tomlin, 1994). NA - not available.

Name	oral LD_{50} /mg/kg (rats)	oral LD_{50} /mg/kg (other species)
Atrazine	1900-3000	750 (rabbits)
Simazine	>5000	500-1400 (sheep)
Cyanazine	180-380	NA
Terbutylazine	1000-1590	NA
Atraton	1465-2400	NA
Terbumeton	>650	NA
Ametryn	110-1750	NA
Prometryn	3150-5235	NA
Terbutryn	2000-2980	3880 (mice)

Table 2. Acute toxicity data for some triazines (IUPAC Agrochemical Information, 2011; Stevens & Sumner, 1991). NA - not available.

However, the situation is less plausible when assessing the chronic toxicity of triazines. Significant scientific and public controversy has been increasing in the last decade especially regarding the effects of environmentally relevant concentrations of atrazine and its main transformation products desethylatrazine, desisopropylatrazine and hydroxyatrazine, resulting in the 2003 ban of atrazine products in European Union (Sass & Colangelo, 2006). Initial studies reported some carcinogenic, mutagenic and teratogenic effects of triazines only at the dose exceeding the maximal tolerable dose (Stevens & Sumner, 1991). However, environmentally relevant low concentrations of atrazine were later shown to adversely affect the normal male development in amphibians (Tavera-Mendoza et al., 2002), although the evidence is still not conclusive (Solomon et al., 2008). Adverse effects of atrazine were shown also for rats, both on male reproductive tract (Kniewald et al., 2000) and on oestrus in females (Eldridge et al., 1999). The latter is presumably due to the effect on hypothalamic-pituitary-gonadal axis and not on intrinsic estrogenic effect of atrazine (Eldridge et al., 1999; Taketa et al., 2011). Similar effects have been observed for the main atrazine transformation products (Stanko et al., 2010). Besides these endocrine-disrupting properties, atrazine has been shown to affect immune function in mice and the effects persist some time after the exposure (Filipov et al., 2005). Other triazine herbicides are not that extensively covered regarding their chronic toxicity, presumably because they are less widely applied. However, USA Environmental Protection Agency (EPA) concludes that triazines and TPs with chlorine attached to the ring (see Fig. 1) have the same common mechanism of toxicity regarding their endocrine-related developmental, reproductive and carcinogenic effects (Environmental Protection Agency [EPA], 2011).

4. Distribution of triazines in the environmental compartments

After the introduction in the environment, triazines are distributed between the three main environmental compartments, namely gaseous (air), aqueous (ground and surface waters) and solid (soil, sediments). The fourth important compartment interacting with the environment is biota: uptake of triazines into microorganisms and plants, which will be considered separately. Distribution is governed by the physicochemical properties of the compounds (Table 1) and is an ongoing process. There is a dynamic interchange of temporary equilibrium states and re-distribution, influenced by weather conditions, input of materials and various pollutants into the environment etc.

Volatilization of triazines and their long-range atmospheric transport is a poorly researched process. It is supposed that, similar to other semivolatiles, triazines are transported by air masses absorbed on the particulate matter and deposit in cold atmospheric conditions (high mountains, higher geographical latitudes) mainly by wet deposition. Snow is an effective scavenger of particulate matter and associated pollutants from the atmosphere. Triazines have been detected both in snow and rainwater (Polkowska et al., 2000; Usenko et al., 2005).

Triazines are distributed mainly between aqueous and solids compartments. The main processes are partitioning and sorption on solid components, as well as solubilisation in the aqueous compartment followed by leaching into lower solid layers and eventually into groundwater. Living organisms present in both compartments contribute to the transport by uptaking the compounds and returning them mainly as transformation products. The majority of research has been done on atrazine in the last decade of 20th century and is encompassed in a recent review paper (Mudhoo & Garg, 2011). However, atrazine residues in soil have proven to be more persistent than previously expected (Jablonowski et al., 2011) and thus there is an ongoing need for further research on the soil behaviour of this compound (Barton & Karathanasis, 2003; Jablonowski et al., 2011; Kovaios et al., 2006; Ling et al., 2006). Atrazine is expected to be in its non-ionized form at the environmentally relevant pH values (see Table 1) and for uncharged compounds, it is generally accepted that they are sorbed on organic carbon fraction of the soils/sediments (Mudhoo & Garg, 2011). The main mechanism in operation is partitioning between aqueous and organic carbon phase, predominantly humic substances. Both overall partition coefficient K_d and partition coefficient for organic carbon K_{oc} are used to quantitatively express the extent of interaction. The reported values for the latter differ considerably from 25 to 600 L/kg OC (Mudhoo & Garg, 2011), which may reflect the differences in organic matter structure. Humic substances (HS) are heterogeneous and still poorly characterized macromolecules or supramolecular associations (Schaumann, 2006). A number of mechanisms have been proposed for the interaction of atrazine and HS: partitioning resulting from hydrophobic interactions (Lima et al., 2010; Prosen & Zupančič-Kralj, 2000), hydrogen bonding (Prosen & Zupančič-Kralj, 2000), electron transfer, charge transfer (Mudhoo & Garg, 2011). While atrazine is sorbed primarily onto soil organic matter (SOM), presence or addition of dissolved organic matter (DOM) may enhance the sorption at lower DOM concentration, but decrease it at higher DOM concentration (Ling et al., 2006; Mudhoo & Garg, 2011), which is a consequence of increased solubilisation of atrazine in the aqueous fraction with DOM.

Atrazine is sorbed on some mineral components of soils/sediments as well: aluminium-saturated smectite (Mudhoo & Garg, 2011), silicagel (Kovaios et al., 2006) and Florisil (SiO_2+MgO) (Prosen et al., 2007), but not calcite or alumina (Kovaios et al., 2006; Prosen et al., 2007). The proposed mechanism is electrostatic or electron-transfer interaction of atrazine with silanol groups (Kovaios et al., 2006; Prosen et al., 2007). Besides soil organic matter (SOM) content and presence of adsorbing minerals, other parameters govern the extent of atrazine sorption to environmental solids: pH, ionic strength, surface area, particle and pore size, presence of other compounds, especially surfactants (J.F. Lee et al., 2004), temperature (Mudhoo & Garg, 2011). Contact time is another important factor. Desorption hysteresis has been observed for longer contact times (Drori et al., 2005; Prosen & Zupančič-Kralj, 2000). The currently accepted model explaining the effect of contact time, nonlinear sorption kinetics, desorption hysteresis and conditioning effect of sorbate on sorbent affinity is the dual-mode sorption process of sorbate in the interchangeable rubbery and glassy state of polymerous SOM material (Schaumann, 2006).

Leaching of atrazine into lower layers of the soil and eventually groundwater is generally affected by the same parameters as sorption. The mobility of compound in soil/sediment is expressed by retardation factor R_f as determined by column lysimeters (Weber et al., 2007). For atrazine, R_f has been shown to be inversely proportional to SOM content and related to pH and soil leaching potential (Weber et al., 2007). Presence of more polar SOM with higher ratio of polar functional groups, e.g. from the manure, has been postulated to result in stronger hydrogen bonding of atrazine and reduced desorption and mobility (Lima et al., 2010), although completely opposite results, i.e. stronger bonding to more hydrophobic humic matter, were reported elsewhere (Celano et al., 2008). Desorption and leaching is enhanced by the presence of surfactants, especially anionic (J.F. Lee et al., 2004; Ying et al., 2005), as well as dissolved organic matter (DOM) (Ling et al., 2006). However, great caution is needed when extrapolating results from these studies to predict the dissipation behaviour of atrazine, as gross underestimations have been observed (Jablonowski et al., 2011).

Considerably less information about sorption and mobility in soil and sediments is available for other triazines or transformation products. Chlorotriazines are generally assumed to behave similarly to atrazine and this has been confirmed in some experiments for simazine (Mudhoo & Garg, 2011; Ying et al., 2005) or terbutylazine for humic organic matter (Celano et al., 2008). The latter is a less polar compared to atrazine and has been shown to exhibit greater extent of sorption on HS (Erny et al., 2011; Prosen & Zupančič-Kralj, 2000). In comparison of methylthio-, methoxy- and chlorotriazine sorption on sediments and mineral soil components, sorption intensity was related to the basicity (pK_a) and water solubility of compounds, but not their $\log K_{ow}$ (Prosen et al., 2007; Stipičević et al., 2009) - Fig. 2. Dealkylated triazine transformation products are weakly sorbed on humic substances compared to parent compounds (Erny et al., 2011), while hydroxyatrazine, a dechlorinated atrazine TP, is extensively sorbed on mineral components of the soil/sediment (Stipičević et al., 2009).

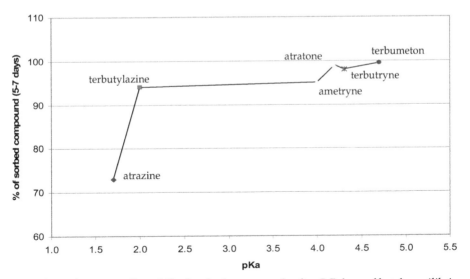

Fig. 2. Relation between pK_a and % of sorbed compounds after 5-7 days of batch equilibrium experiment on Florisil (SiO$_2$, MgO). Adapted after Prosen et al. (2007).

Knowledge of sorption/desorption behaviour of triazines is frequently applied in bioremediation either to enhance their leaching or to stabilise the residues in the contaminated sites (Delgado-Moreno et al., 2010; Jones at al., 2011; J.F. Lee et al., 2004; Lima et al., 2010; Mudhoo & Garg, 2011; Ying et al., 2005).

5. Triazine degradation and uptake in the soil

The sorption behaviour of triazines in soil directly influences their bioavailability to soil microorganisms and plants (Mudhoo & Garg, 2011), leading to their uptake and biodegradation. Numerous studies are available for atrazine as the most widely applied and apparently also persistent triazine in the soil (Jablonowski et al., 2011). Plant uptake of triazines from the contaminated soils is extensively studied as a means for bioremediation. The C4-metabolism plants show the greatest resistance to triazines and detoxify them by hydrolysis. Examples of plants shown to be useful in degrading atrazine in their rhizosphere are *Polygonum lapathifolium, Panicum dichotomiflorum* (Mudhoo & Garg, 2011), *Pennisetum clandestinum* (Popov & Cornish, 2006; Singh et al., 2004).

Ongoing research in the soil microorganisms capable of utilizing triazines as their energy source has resulted in an extensive array of isolated strains: *Acinetobacter* sp., *Cytophaga* sp., *Pseudomonas* sp., *Ralstonia* sp., *Agrobacterium* sp. (Mudhoo & Garg, 2011), *Klebsiella* sp. and *Comamonas* sp. (Yang et al., 2010), *Nocardioides* sp. and *Arthrobacter* sp. (Vibber et al., 2007). Most of them are capable of extensive mineralization of triazines (Mudhoo & Garg, 2011; Yang et al., 2010) and have a limited access even to aged herbicide residues in the soil (Jablonowski et al., 2008; Mudhoo & Garg, 2011). The species most often used for triazine degradation is *Pseudomonas* sp., its efficacy has been shown to be influenced by citrate addition (Jablonowski et al., 2008), soil humidity (Ngigi et al., 2011) and microorganism adsorption on simulated soil particle aggregates (Alekseeva et al., 2011). Green algae and diatoms (Mudhoo & Garg, 2011), as well as cyanobacteria (Gonzalez-Barreiro et al., 2006) are also capable of atrazine uptake and are thus a valuable option for the bioremediation of the contaminated waters. Certain fungal species able to grow on atrazine-contaminated soils and capable of its uptake have been identified as well (Mudhoo & Garg, 2011).

Compared to biotic degradation by microorganisms and higher plants, abiotic degradation of triazines in soils is a minor dissipation route. Humic substances at low pH catalyse the hydrolysis of atrazine and its chlorinated transformation products to their hydroxy analogues (Prosen & Zupančič-Kralj, 2005). Photolysis of atrazine under solar irradiation and in the presence of humic substances was found to be negligible (Prosen & Zupančič-Kralj, 2005); however, simazine and terbutylazine were found to dissipate faster under solar irradiation of the soil (Navarro et al., 2009). Photolytic transformation and eventual mineralization is enhanced by using a suitable photocatalytic agent, e.g. TiO_2, which holds a potential for clean-up of contaminated sites (Konstantinou & Albanis, 2003).

6. Analytical approaches and cautions for triazine determination in soil

Triazine determination data for soil and other solid environmental samples are used to estimate the extent of the site pollution and potential toxicity (Jablonowski et al., 2011). However, determination of triazines and their TPs in solid samples is prone to many problems, as schematically depicted in Fig. 3.

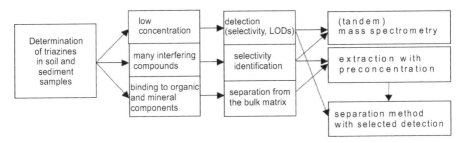

Fig. 3. Schematic representation of the problems and solutions for triazine determination in solid environmental samples.

Technique	Principle	Advantages	Disadvantages
Soxhlet Extraction (SE)	continuous percolation of organic solvent	- recoveries not dependent on sample type - cheap	- time-consuming - high consumption of organic solvents - extracts have to be concentrated
Ultrasonication Extraction (USE)	mixing, desorption of analytes from sample components	- recoveries not dependent on sample type - cheap	- moderately time-consuming - high consumption of organic solvents - work-intensive
Supercritical Fluid Extraction (SFE)	supercritical fluid of low viscosity better penetrates the sample	- fast method - solvent CO_2 non-toxic, environmentally acceptable	- limited sample amount - recoveries depend on sample type - high initial cost
Microwave-Assisted Solvent Extr. (MASE)	microwave- assisted desorption of analytes and sample components	- fast method - low consumption of organic solvents - additional regulation parameters	- polar solvents only - non-selective - extensive extract clean-up needed - high initial cost
Pressurised Liquid Extraction (PLE) / Accelerated Solvent Extr. (ASE)	enhanced extraction efficiency of analytes due to solvents at high temperature and pressure (liquids above boiling point)	- fast method - low consumption of organic solvents	- non-selective - extensive extract clean-up needed - high initial cost

Table 3. Common extraction techniques for triazines from the solid environmental samples (Andreu & Pico, 2004; Camel, 2000; Lesueur et al., 2008; Lopez-Avila, 1999).

The analytical procedure usually comprises of a suitable extraction technique (Table 3), preferably enabling preconcentration as well, possibly a clean-up step, and an appropriate determination technique (Andreu & Pico, 2004; Camel, 2000; Lesueur et al., 2008; Lopez-Avila, 1999). The first dilemma encountered is whether to use an exhaustive extraction technique or a more mild one. Extraction techniques regarded as exhaustive under most conditions are Soxhlet's, MASE and PLE (Camel, 2000). There is a high probability that even triazines bound to soil components would be extracted, although this may depend on the type of compound (Kovačić et al., 2004). However, most of the unwanted organic compounds from the sample would be transferred to extract as well, and these interferences have to be selectively removed prior to analysis by an appropriate clean-up technique. The key word in this case is selectivity, as the clean-up may otherwise lead to significant loss of analytes as well. A selection of frequently applied clean-up techniques is listed in Table 4. In the second case, i.e. by applying a mild extraction technique (0.01 M CaCl$_2$ solution or aqueous methanol), the obtained extract would better reflect the actual fraction of the triazines and TPs available to plants and microorganisms (Regitano et al., 2006) and could thus be more useful for the actual assessment of the residual toxicity of triazines (Jablonowski et al., 2008; Jablonowski et al., 2011).

Technique	Principle	Advantages	Disadvantages	Variants and improvements
Liquid-Liquid Extraction (LLE)	partitioning between two immiscible solvents	- high recoveries - broad choice of solvents	- time-consuming - automatisation difficult - environmentally problematic	supported liquid membrane extr. (SLME) liquid-phase microextr. (LPME) / single-drop microextraction
Solid Phase Extraction (SPE)	adsorption / partitioning between aqueous and solid phase, followed by desorption with organic solvents	- high recoveries - low solvent consumption - automatisation possible (on-line)	- lower selectivity - narrower choice of sorbents	restricted access material (RAM) molecularly imprinted polymer (MIP) immunosorbents multi-walled nanotubes
Solid Phase Micro-extraction (SPME)	partitioning between aqueous and non-polar phase on fibre, follow by thermal or solvent desorption	- fast - solventless - automatisation possible	- mainly for volatile compounds - poor repeatability - non-exhaustive (low recoveries)	in-tube SPME

Table 4. Common clean-up techniques for triazines in soil/sediment extracts (Andreu & Pico, 2004; Hylton & Mitra, 2007; Jonsson & Mathiasson, 2000; Masque et al., 2001; Min et al., 2008; Psillakis & Kalogerakis, 2002; Stalikas et al., 2002).

Determination of triazines in the extracts after extraction and clean-up is usually accomplished using either gas (GC) or liquid chromatography (HPLC) (Andreu & Pico, 2004). Both techniques can be coupled with mass spectrometry, enabling simultaneous confirmation of compound identity (Andreu & Pico, 2004; Lesueur et al., 2008; Min et al., 2008; Tsang et al., 2009; Usenko et al., 2005). Other detectors frequently used in triazine analysis are spectrophotometric, preferably diode-array detector for HPLC (Andreu & Pico, 2004; Kovačić et al., 2004; Prosen et al., 2004), and nitrogen-phosphorous detector for GC (Andreu & Pico, 2004; Stalikas et al., 2002).

Besides chromatography, other analytical techniques are seldom applied to triazine determination, although they may offer some significant advantages: electromigration techniques, e.g. micellar electrokinetic chromatography (Lima et al., 2009; Prosen et al., 2004); voltammetry (De Souza et al., 2007). Biosensors and bioassays are used for preliminary screening of samples or sample extracts, but because of their cross-reactivity the samples with analyte content above the cut-off value should be re-analysed by a more specific analytical technique. The most widely applied is antibody-based ELISA, but some innovative approaches have been developed, e.g. sensors based on photosystem-II inhibition from plant photosynthetic membranes (Bengtson Nash et al., 2005; Varsamis et al., 2008).

Analytical determination of triazines in solid samples, although often seen as a routine procedure, is prone to many errors. Starting with sampling, the sample taken for analysis should be representative of that part of environment for which the information about pollutant concentration should be obtained. To achieve this goal, an appropriate number of samples, as well as time and site of sampling should be considered. Preservation of samples during the transport and storage is important as well and should be carefully selected (Kebbekus & Mitra, 1998). An example is the need to completely dechlorinate drinking water to prevent rapid degradation of triazines (Smith et al., 2008). Next step, namely extraction with clean-up, is again critical due to the possibility of significant analyte losses because of improper sample preparation conditions. These should be optimised and tested for every analyte. The choice between exhaustive and milder extraction techniques has already been mentioned, but mild conditions are also needed to avoid thermal degradation. Most triazines and their TPs are thermally stable, but not all (Tsang et al., 2009). Another caveat with extraction is the significant difference in analyte binding and thus extraction recoveries between freshly-spiked blank samples and real-life samples containing the so-called »aged residues«. Various authors have proposed to reproduce aging under environmental conditions by leaving spiked blank samples at room temperature for anything between 3 days and 2 years (Andreu & Pico, 2004). However, simulation may not necessarily yield equivalent results to field conditions (Louchart & Voltz, 2007). Finally, determination technique is important in terms of selectivity, limits of detection and reliable quantification. To achieve the latter, standard solutions for the calibration should always match the actual matrix as close as possible to avoid the significant matrix effects seen with some types of detectors (Kovačić et al., 2004), especially with electrospray interface for LC-MS.

7. Elucidation of triazine fate in the soil as influenced by analytical determination

As already explored in subchapter 4 of this review, we are mainly concerned with triazine sorption, desorption, leaching and plant/microorganism uptake when dealing with triazine fate in the soil. Sorption in its broadest sense (i.e. partitioning, non-covalent and covalent binding) is usually evaluated by sorption isotherms conforming to various theoretical models: Freundlich, Langmuir, Polanyi-Dubinin-Manes, etc. (Aboul-Kasim & Simoneit, 2001; Kleineidam et al., 2002). The most frequently used method to obtain the experimental data for isotherm construction remains the batch equilibrium method (Celano et al., 2008; Kleineidam et al., 2002; Konda et al., 2002; Kovaios et al., 2006; Lima et al., 2009; Ling et al., 2006; Stipičević et al., 2009). Other approaches are by chromatographic estimation (Bermudez-Saldana et al., 2006) or indirectly by structural descriptors (Schüürmann et al., 2006). In batch equilibrium method, several variables may influence the process of sorption and have to be carefully optimised: organic solvent content, ionic strength and pH, solid/solution ratio, sorption time (Celano et al., 2008; Kleineidam et al., 2002; Kovaios et al., 2006; Prosen & Zupančič-Kralj, 2000; Prosen et al., 2007). After the equilibrium is reached, the solution has to be separated from the sorbent either by centrifugation or filtration (Kleineidam et al., 2002). By using the latter, another potential source of error is introduced as more hydrophobic compounds may bind to certain types of filters.

The equilibrium concentration of the pollutant in the solution after the separation is determined by any of the analytical methods mentioned in subchapter 6. Preferably, it should be performed without previous extraction as this introduces another equilibrium and another possible source of error. Thus, direct HPLC (Celano et al., 2008; Prosen & Zupančič-Kralj, 2000) or electromigration techniques (Erny et al., 2011; Lima et al., 2009) are the methods of choice. If radiolabelled compounds are used, their equilibrium concentration can be measured by radioactivity measurement (Jablonowski et al., 2008). A different approach is to determine the free concentration directly in a multi-phase system by a non-exhaustive solid-phase microextraction and subsequent GC analysis (Heringa & Hermens, 2003; S. Lee et al., 2003; Prosen et al., 2007). The depletion of the compounds from the solution is considered to be negligible, thus giving the opportunity to measure the true equilibrium concentration in the solution (Heringa & Hermens, 2003). Distribution coefficients K_d obtained by SPME-GC determination of equilibrium concentrations after the sorption experiment have been reported to be significantly different compared to those obtained by other determination methods (S. Lee et al., 2003).

As well as for sorption/desorption, the understanding of the leaching behaviour of triazines is significantly influenced by the determination method. The usual approach is to evaluate the mobility of the compound in soil columns by lysimeters (Jablonowski et al., 2011; Weber et al., 2007), but experiments should be conducted under the appropriate time-scale to avoid gross underestimations (Jablonowski et al., 2011). A different approach is the use of ceramic suction cups, but these are also prone to errors due to ageing effects (Domange et al., 2004).

8. Conclusions

This review attempts to cover a vast subject of triazine behaviour in the environment, especially soil, as well as their analytical determination in the same. Special attention was given to the various problems encountered in both. However, the broadness of the subject prevents its detailed evaluation; the interested reader can find more information in other excellent reviews that focus more on triazine behaviour in solid environmental compartment (Jablonowski et al., 2011; Mudhoo & Garg, 2011), their degradation and elimination (Konstantinou & Albanis, 2003) or the applied analytical methods (Andreu &

Pico, 2004; Camel, 2000; Hylton & Mitra, 2007; Jonsson & Mathiasson, 2000; Lopez-Avila, 1999; Masque et al., 2001; Psillakis & Kalogerakis, 2002).

9. Acknowledgment

Author would like to acknowledge the financial support from the Ministry of Higher Education, Science and Technology of the Republic Slovenia through Grant P1-0153.

10. References

Aboul-Kasim, T. A. T., Simoneit, B. R. T. (2001). *Pollutant-solid phase interactions.* Springer, Berlin, Germany.

Alekseeva, T., Prevot, V., Sancelme, M., Forano, C., Besse-Hoggan, P. (2011). Enhancing atrazine biodegradation by *Pseudomonas* sp. strain ADP adsorption to Layered Double Hydroxide bionanocomposites. *Journal of Hazardous Materials*, Vol.191, 126-135.

Andreu, V., Picó, Y. (2004). Determination of pesticides and their degradation products in soil: critical review and comparison of methods. *Trends in Analytical Chemistry*, Vol.23, 772-789.

Barton, C. D., Karathanasis, A. D. (2003). Influence of soil colloids on the migration of atrazine and zinc through large soil monoliths. *Water, Air and Soil Pollution*, Vol.143, 3-21.

Bengtson Nash, S. M., Schreiber, U., Ralph, P. J., Müller, J. F. (2005). The combined SPE:ToxY-PAM phytotoxicity assay; application and appraisal of a novel biomonitoring tool for the aquatic environment. *Biosensors Bioelectronics*, Vol.20, 1443-1451.

Bermúdez-Saldaña, J. M., Escuder-Gilabert, L., Medina-Hernández, M. J., Villanueva-Camañas, R. M., Sagrado, S. (2006). Chromatographic estimation of the soil-sorption coefficients of organic compounds. *Trends in Analytical Chemistry*, Vol.25, 122-132.

Camel, V. (2000). Microwave-assisted solvent extraction of environmental samples. *Trends in Analytical Chemistry*, Vol.19, 229-248.

Celano, G., Šmejkalova, D., Spaccini, R., Piccolo, A. (2008). Interactions of three *s*-triazines with humic acids of different structure. *Journal of Agricultural Food Chemistry*, Vol.56, 7360-7366.

Delgado-Moreno, L., Pena, A., Almendros, G. (2010). Contribution by different organic fractions to triazines sorption in Calcaric Regosol amended with raw and biotransformed olive cake. *Journal of Hazardous Materials*, Vol.174, 93-99.

De Souza, D., de Toledo, R. A., Galli, A., Salazar-Banda, G. R., Silva, M. R. C., Garbellini, G. S., Mazo, L. H., Avaca, L. A., Machado, S. A. S. (2007). Determination of triazine herbicides: development of an electroanalytical method utilizing a solid amalgam electrode that minimizes toxic waste residues, and a comparative study between voltammetric and chromatographic techniques. *Analytical Bioanalytical Chemistry*, Vol.387, 2245-2253.

Domange, N., Grégoire, C., Gouy, V., Tremolières, M. (2004). Effet du vieillissement des céramiques poreuses sur leur capacité à évaluer la concentration de pesticide en solution. Abridged English version. *C. R. Geoscience*, Vol.336, 49-58.

Drori, Y., Aizenshtat, Z., Chefetz, B. (2005). Sorption-desorption behavior of atrazine in soils irrigated with reclaimed wastewater. *Soil Science Society of America Journal*, Vol.69, 1703-1710.

Eldridge, J. C., Wetzel, L. T., Tyrey, L. (1999). Estrous cycle patterns of Sprague-Dawley rats during acute and chronic atrazine administration. *Reproductive Toxicology*, Vol.13, 491-499.

Environmental Protection Agency (2006). Cumulative Risk From Triazine Pesticides. Doc. ID EPA-HQ-OPP-2005-0481-0003. Available from: http://www.regulations.gov/

Erny, G. L., Calisto, V., Lima, D. L. D., Esteves, V. I. (2011). Studying the interaction between triazines and humic substances—A new approach using open tubular capillary eletrochromatography. *Talanta*, Vol.84, 424-429.

Filipov, N. M., Pinchuk, L. M., Boyd, B. L., Crittenden, P. L. (2005). Immunotoxic effects of short-term atrazine exposure in young male C57BL/6 mice. *Toxicological Science*, Vol.86, 324-332.

Gonzalez-Barreiro, O., Rioboo, C., Herrero, C., Cid, A. (2006). Removal of triazine herbicides from freshwater systems using photosynthetic microorganisms. *Environmental Pollution*, Vol.144, 266-271.

Heringa, M. B., Hermens, J. L. M. (2003). Measurement of free concentrations using negligible depletion-solid phase microextraction (nd-SPME). *Trends in Analytical Chemistry*, Vol.22, 575-587.

Hylton, K., Mitra, S. (2007). Automated, on-line membrane extraction. Review. *Journal of Chromatography A*, Vol. 152, 199-214.

IUPAC Agrochemical Information (2011). Available from: http://sitem.herts.ac.uk/aeru/iupac/

Jablonowski, N. D., Modler, J., Schaeffer, A., Burauel, P. (2008). Bioaccessibility of environmentally aged ^{14}C-atrazine residues in an agriculturally used soil and its particle-size aggregates. *Environmental Science Technology*, Vol.42, 5904-5910.

Jablonowski, N. D., Schäffer, A., Burauel, P. (2011). Still present after all these years: persistence plus potential toxicity raise questions about the use of atrazine. *Environmental Science of Pollution Research*, Vol.18, 328-331.

Jones, D. L., Edwards-Jones, G., Murphy, D. V. (2011). Biochar mediated alterations in herbicide breakdown and leaching in soil. *Soil Biology Biochemistry*, Vol.43, 804-813.

Jönsson, J. Å., Mathiasson, L. (2000). Membrane-based techniques for sample enrichment. Review. *Journal of Chromatography A*, Vol.902, 205-225.

Kaune, A., Brüggemann R., Kettrup, A. (1998). High-performance liquid chromatographic measurement of the 1-octanol-water part. coefficient of s-triazine herbicides and some of their degradation products. *Journal of Chromatography A*, Vol.805, 119-126.

Kebbekus, B. B., Mitra, S. (1998). *Environmental chemical analysis*. Chapman & Hall/CRC, Boca Raton, FL, USA.

Kleineidam, S., Schueth, C., Grathwohl, P. (2002). Solubility-normalized combined adsorption-partitioning sorption isotherms for organic pollutants. *Environmental Science Technology*, Vol.36, 4689-4697.

Kniewald, J., Jakominić, M., Tomljenović, A., Šimić, B., Romac, P., Vranešić, Đ., Kniewald, Z. (2000). Disorders of male rat reproductive tract under the influence of atrazine. *Journal of Applied Toxicology*, Vol.20, 61-68.

Konda, L. N., Czinkota, I., Fueleky, G., Morovjan, G. (2002). Modeling of single-step and multistep adsorption isotherms of organic pesticides on soil. *Journal of Agricultural Food Chemistry*, Vol.50, 7326-7331.

Konstantinou, I. K., Albanis, T. A. (2003). Photocatalytic transformation of pesticides in aqueous TiO$_2$ suspensions using artificial and solar light: intermediates and degradation pathways. Review. *Applied Catalysis B: Environmental*, Vol.42, 319-335.

Kovačić, N., Prosen, H., Zupančič-Kralj, L. (2004). Determination of triazines and atrazine metabolites in soil by microwave-assisted solvent extraction and high-pressure liquid chromatography with photo-diode-array detection. *Acta Chimica Slovenica*, Vol.51, 395-407.

Kovaios, I. D., Paraskeva, C. A., Koutsoukos, P. G., Payatakes, A. C. (2006). Adsorption of atrazine on soils: Model study. *Journal of Colloid Interface Science*, Vol.299, 88-94.

Lee, J. F., Hsu, M. H., Chao, H. P., Huang, H. C., Wang, S. P. (2004). The effect of surfactants on the distribution of organic compounds in the soil solid/water system. *Journal of Hazardus Materials B*, Vol.114, 123-130.

Lee, S., Gan, J., Liu, W. P., Anderson, M. A. (2003). Evaluation of K_d underestimation using solid phase microextraction. *Environmental Science Technology*, Vol.37, 5597-5602.

Lesueur, C., Gartner, M., Mentler, A., Fuerhacker, M. (2008). Comparison of four extraction methods for the analysis of 24 pesticides in soil samples with gas chromatography–mass spectrometry and liquid chromatography–ion trap–mass spectrometry. *Talanta*, Vol.75, 284-293.

Lima, D. L. D., Erny, G. L., Esteves, V. I. (2009). Application of MEKC to the monitoring of atrazine sorption behaviour on soils. *Journal of Separation Science*, Vol.32, 4241-4246.

Lima, D. L. D., Schneider, R. J., Scherer, H. W., Duarte, A. C., Santos, E. B. H., Esteves, V. I. (2010). Sorption-desorption behavior of atrazine on soils subjected to different organic long-term amendments. *Journal of Agricultural Food Chemistry*, Vol.58, 3101-3106.

Ling, W., Xu, J., Gao, Y. (2006). Dissolved organic matter enhances the sorption of atrazine by soil. *Biological Fertilization of Soils*, Vol.42, 418-425.

Lopez-Avila, V. (1999). Sample preparation for environmental analysis. *Critical Reviews in Analytical Chemistry*, Vol.29, 195-230.

Louchart, X., Voltz, M. (2007). Aging effects on the availability of herbicides to runoff transfer. *Environmental Science Technology*, Vol.41, 1137-1144.

Masqué, N., Marcé, R. M., Borrull, F. (2001). Molecularly imprinted polymers: new tailor-made materials for selective solid-phase extraction. *Trends in Analytical Chemistry*, Vol.20, 477-486.

Min, G., Wang, S., Zhu, H., Fang, G., Zhang, Y. (2008). Multi-walled carbon nanotubes as solid-phase extraction adsorbents for determination of atrazine and its principal metabolites in water and soil samples by gas chromatography-mass spectrometry. *Science of the Total Environment*, Vol.396, 79-85.

Mudhoo, A., Garg, V.K. (2011). Sorption, transport and transformation of atrazine in soils, minerals and composts: a review. *Pedosphere*, Vol.21, 11-25.

Navarro, S., Bermejo, S., Vela, N., Hernandez, J. (2009). Rate of loss of simazine, terbuthylazine, isoproturon, and methabenzthiazuron during soil solarization. *Journal of Agricultural Food Chemistry*, Vol.57, 6375-6382.

Ngigi, A., Dörfler, U., Scherb, H., Getenga, Z., Boga, H., Schroll, R. (2011). Effect of fluctuating soil humidity on in situ bioavailability and degradation of atrazine. *Chemosphere*, Vol.84, 369-375.

Noble, A. (1993). Partition coefficients (*n*-octanol-water) for pesticides. *Journal of Chromatography*, Vol.642, 3-14.

Polkowska, Ž., Kot, A., Wiergowski, M., Wolska, L., Wolowska, K., Namiesnik, J. (2000). Organic pollutants in precipitation: determination of pesticides and polycyclic aromatic hydrocarbons in Gdansk, Poland. *Atmospheric Environment*, Vol.34, 1233-1245.

Popov, V. H., Cornish, P.S. (2006). Atrazine tolerance of grass species with potential for use in vegetated filters in Australia. *Plant Soil*, Vol.280, 115-126.

Prosen, H., Zupančič-Kralj, L. (2000). The interaction of triazine herbicides with humic acids. *Chromatographia Supplement*, Vol. 51, S155-S164.

Prosen, H., Guček, M., Zupančič-Kralj, L. (2004). Optimization of liquid chromatography and micellar electrokinetic chromatography for the determination of atrazine and its first degradation products in humic waters without sample preparation. *Chromatographia Supplement*, Vol.60, 107-112.

Prosen, H., Zupančič-Kralj, L. (2005). Evaluation of photolysis and hydrolysis of atrazine and its first degradation products in the presence of humic acids. *Environmental Pollution*, Vol.133, 517-529.

Prosen, H., Fingler, S., Zupančič-Kralj, L., Drevenkar, V. (2007). Partitioning of selected environmental pollutants into organic matter as determined by solid-phase microextraction. *Chemosphere*, Vol.66, 1580-1589.

Psillakis, E., Kalogerakis, N. (2002). Developments in liquid-phase microextraction. *Trends in Analytical Chemistry*, Vol.21, 53-63.

Regitano, J. B., Koskinen, W. C., Sadowsky, M. J. (2006). Influence of soil aging on sorption and bioavailability of simazine. *Journal of Agricultural Food Chemistry*, Vol.54, 1373-1379.

Sass, J. B., Colangelo, A. (2006). European Union bans atrazine, while the United States negotiates continued use. *International Journal of Occupational and Environmental Health*, Vol.12, 260-267.

Schaumann, G. E. (2006). Review Article - Soil organic matter beyond molecular structure. Part I: Macromolecular and supramolecular characteristics. *Journal of Plant Nutrition and Soil Science*, Vol.169, 145-156.

Schüürmann, G., Ebert, R. U., Kühne, R. (2006). Prediction of sorption of organic compounds into soil organic matter from molecular structure. *Environmental Science Technology*, Vol.40, 7005-7011.

Shiu, W.Y., Ma, K.C., Mackay, D., Seiber, J.N., Wauchope, R.D. (1990). Solubilities of pesticide chemicals in water. Part II: data compilation. *Reviews of. Environmental Contamination and Toxicology*, Vol.116, 15-187.

Singh, N., Megharaj, M., Kookana, R. S., Naidu, R., Sethunathan, N. (2004). Atrazine and simazine degradation in Pennisetum rhizosphere. *Chemosphere*, Vol.56, 257-263.

Smith, G. A., Pepich, B. V., Munch, D. J. (2008). Preservation and analytical procedures for the analysis of chloro-*s*-triazines and their chlorodegradate products in drinking waters using direct injection liquid chromatography tandem mass spectrometry. *Journal of Chromatography A*, Vol.1202, 138-144.

Solomon, K. R., Carr, J. A., Du Preez, L. H., Giesy, J. P., Kendall, R. J., Smith, E. E., Van Der Kraak, G. J. (2008). Effects of atrazine on fish, amphibians, and aquatic reptiles: A critical review. *Critical Reviews in Toxicology*, Vol.38, 721-772.

Stalikas, C., Knopp, D., Niessner, R. (2002). Sol-gel glass immunosorbent-based determination of s-triazines in water and soil samples using gas chromatography with a nitrogen phosphorus detection system. *Environmental Science Technology*, Vol.36, 3372-3377.

Stanko, J. P., Enoch, R. R., Rayner, J. L., Davis, C. C., Wolf, D. C., Malarkey, D. E., Fenton, S. E. (2010). Effects of prenatal exposure to a low dose atrazine metabolite mixture on pubertal timing and prostate development of male Long-Evans rats. *Reproductive Toxicology*, Vol.30, 540-549.

Stevens, J. T., Sumner, D. D. (1991). Herbicides. In: *Handbook of Pesticide Toxicology*, Hayes, W. J., Laws, E.R. (eds.), Academic Press, San Diego, CF, USA.

Stipičević, S., Fingler, S., Drevenkar, V. (2009). Effect of organic and mineral soil fractions on sorption behaviour of chlorophenol and triazine micropollutants. *Archives of Hygiene and Occupational Toxicology*, Vol.60, 43-52.

Taketa, Y., Yoshida, M., Inoue, K., Takahashi, M., Sakamoto, Y., Watanabe, G., Taya, K., Yamate, J., Nishikawa, A. (2011). Differential stimulation pathways of progesterone secretion from newly formed corpora lutea in rats treated with ethylene glycol monomethyl ether, sulpiride, or atrazine. *Toxicological Sciences*, Vol.121, 267-278.

Tavera-Mendoza, L., Ruby, S., Brousseau, P., Fournier, M., Cyr, D., Marcogliese, D. (2002). Response of the amphibian tadpole (*Xenopus laevis*) to atrazine during sexual differentiation of the testis. *Environmental Toxicology and Chemistry*, Vol.21, 527-531.

Tomlin, C. (ed.) (1994). *The Pesticide Manual Incorporating the Agrochemicals Handbook*. British Crop Protection Council, Surrey, UK.

Tsang, V. W. H., Lei, N. Y., Lam, M. H. W. (2009). Determination of Irgarol-1051 and its related s-triazine species in coastal sediments and mussel tissues by HPLC–ESI-MS/MS. *Marine Pollution Bulletin*, Vol.58, 1462-1471.

Usenko, S., Hageman, K.J., Schmedding, D. W., Wilson, G. R., Simonich, S. L. (2005). Trace analysis of semivolatile organic compounds in large volume samples of snow, lake water, and groundwater. *Environmental Science Technology*, Vol.39, 6006-6015.

Varsamis, D. G., Touloupakis, E., Morlacchi, P., Ghanotakis, D. F., Giardi, M. T., Cullen, D. C. (2008). Development of a photosystem II-based optical microfluidic sensor for herbicide detection. *Talanta*, Vol.77, 42-47.

Vibber, L. L., Pressler, M. J., Colores, G. M. (2007). Isolation and characterization of novel atrazine-degrading microorganisms from an agricultural soil. *Applied Microbiology and Biotechnology*, Vol.75, 921-928.

Weber, J. B., Warren, R. L., Swain, L. R., Yelverton, F. H. (2007). Physicochemical property effects of three herbicides and three soils on herbicide mobility in field lysimeters. *Crop Protection*, Vol.26, 299-311.

Yang, C., Li, Y., Zhang, K., Wang, X., Ma, C., Tang, H., Xu, P. (2010). Atrazine degradation by a simple consortium of *Klebsiella* sp. A1 and *Comamonas* sp. A2 in nitrogen enriched medium. *Biodegradation*, Vol.21, 97-105.

Ying, G.G., Kookana, R. S., Mallavarpu, M. (2005). Release behavior of triazine residues in stabilised contaminated soils. *Environmental Pollution*, Vol.134, 71-77.

Preparation and Characterization of Polymeric Microparticles Used for Controlled Release of Ametryn Herbicide

Fabiana A. Lobo[1], Carina L. de Aguirre[2], Patrícia M.S. Souza[2],
Renato Grillo[2,3], Nathalie F.S. de Melo[2,3],
André H. Rosa[2] and Leonardo F. Fraceto[2]
[1]UFOP - Universidade Federal de Ouro Preto
[2]UNESP – State University of São Paulo,
[3]Department of Environmental Engineering, Campus Sorocaba, SP,
Brazil

1. Introduction

There is increasing pressure to improve agricultural productivity, due to rapid population growth, increased consumption and global demand for high quality products. As a result, agricultural chemicals have become essential for the control of weeds, pests and diseases in a wide range of crops. Ametryn (2-ethylamino-4-isopropylamino-6-methylthio-s-2,4,6-triazine) is a selective herbicide belonging to the s-triazine family, whose activity is the result of inhibition of photosynthesis by blocking of electron transport. The ametryn molecule (Figure 1) contains a symmetrical hexameric aromatic ring in its chemical structure, consisting of three carbon atoms and three nitrogen atoms in alternate positions. The herbicide is classified as a methylthiotriazine, due to the presence of the SCH_3 group (Tennant et al., 2001).

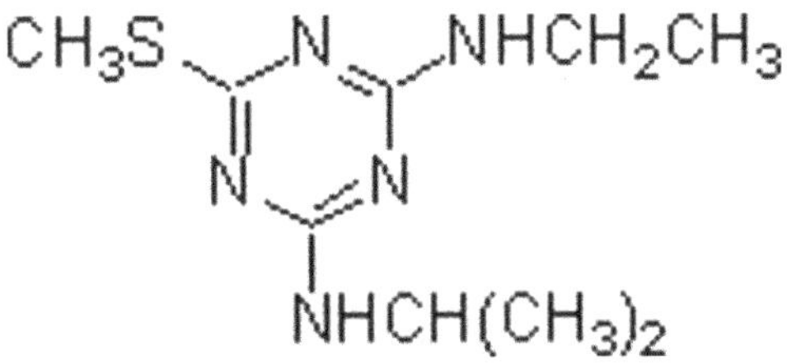

$$CH_3S \diagdown \underset{N \diagdown \diagup N}{N} \diagup NHCH_2CH_3$$

$$NHCH(CH_3)_2$$

Fig. 1. Structural formula of ametryn.

Ametryn is used for the control of graminaceous and broad-leaved weeds in plantations of annual crops (Tennant et al., 2001). Once in the soil the herbicide may be taken up by plants, absorbed by the soil and plant residues, biodegraded, or undergo chemical transformations that increase its volatilization and photocatalytic decomposition. Studies have shown that prolonged human exposure to triazine herbicides can lead to serious health problems including contact dermatitis, intoxication, hormonal dysfunction and cancers (Friedmann et al., 2002). It is therefore desirable to develop techniques whereby the physico-chemical properties of these chemicals can be altered and their usage made safer. The goal is to enable the use of soil management strategies that can produce foods at the current high levels of demand, without significant human or environmental risk.

Micro- and nanostructured polymeric materials can be used as transport systems for active chemicals. Advantages of these materials include good physical, chemical and biological stability, simple and reproducible preparation procedures, and applicability to a wide range of chemicals. In use, the active principle is released slowly and continuously, enabling the use of smaller quantities with greater efficiency, which reduces the risk of adverse environmental impacts (Sinha et al., 2004; Sopena et al., 2009).

Controlled release systems have been extensively used in the food and pharmaceutical industries for active substances including nutrients, drugs and aromas (El Bahri & Taverdet, 2007; Grillo et al., 2008; Mello et al., 2008; Moraes et al., 2010), and there has been a recent increase in their application in medicine (Natarajan et al., 2011; Parajo et al., 2010; Vicente et al., 2010).

Amongst the new controlled-release system technologies under development, the use of polymeric micro- and nanoparticles is of special interest in agribusiness. Several studies have investigated controlled-release systems for bioactive compounds in agricultural applications (Ahmadi & Ahmadi, 2007; Bin Hussein et al., 2010; El Bahri & Taverdet, 2005, 2007; Grillo et al., 2010; Hirech et al., 2003; Li et al., 2010; Lobo et al., 2011; Silva et al., 2010; Singh et al., 2008, 2010). Materials that have been used include silica, bentonite and sepiolite clays, and polymeric substances such as alginate, lignin and synthetic polymers. The latter include the poly(hydroxyalkanoates) (PHAs) (Salehizadeh & Loosdrecht, 2004), of which poly(3-hydroxybutyrate) (PHB) and its hydroxyvalerate copolymer (PHBV) have been most widely used (Amass & Tighe, 1998). The advantages of using polymers such as PHB and PHBV are that they are fully biodegradable, inexpensive and easily prepared by bacterial fermentation (Pouton & Akhtarb 1996; Reis et al., 2008). These polymers are isotactic and highly crystalline (55-80 %), so that their degradation rates are relatively slow compared to those of lactate (PLA) and glycolate (PGA) copolymers (Sudesh et al., 2000).

The objective of this work was to develop a novel release system for ametryn, employing microparticles prepared using two different polymers, PHB and PHBV (either individually or as mixtures). It was envisaged that the encapsulation of the herbicide in these microparticles would improve its chemical stability and enable the use of smaller quantities of the chemical, hence reducing the risk of environmental contamination.

2. Experimental

2.1 Materials

Polyvinyl alcohol (PVA), poly(3-hydroxybutyrate) (PHB, MW = 312,000 g mol^{-1}), poly(3-hydroxybutyrate-co-hydroxyvalerate) (PHBV, MW = 238,000 g mol^{-1}) and ametryn (Pestanal®) were purchased from Sigma Chem. Co. The solvents employed in the chromatographic analyses were acetonitrile, HPLC grade methanol (JT Baker) and Milli-Q water. The solutions were filtered using 0.22 μm nylon membranes (Millipore, Belford, USA).

2.2 Methodology

2.2.1 Determination of ametryn

The HPLC analyses were performed using a Varian ProStar instrument fitted with a PS 210 pump, a UV-VIS detector (PS 325), a Metatherm oven and an automatic injector (PS 410). The chromatograms were acquired and processed using Galaxy Workstation software. The eluent used was acetonitrile/water (70:30, v/v), at a flow rate of 1.4 mL min^{-1}, and separation was achieved using a Phenomenex Gemini C$_{18}$ reversed phase column (5 µm, 110 Å, 150 mm x 4.60 mm i.d.). Ametryn was detected at a wavelength of 260 nm. The injection volume was 100 µL, and all samples were previously filtered through 0.22 µm nylon membranes.

2.2.2 Preparation of the polymeric microparticles containing ametryn

Microparticles were prepared with the PHB and PHBV polymers, used either individually or as a mixture, by formation of oil in water emulsions using the emulsification-solvent evaporation technique (Coimbra et al., 2008; Conti et al., 1995; Lionzo et al., 2007; Lobo et al., 2011). 200 mg of polymer (PHB, PHBV or a mixture of the two polymers, as described in Table 1) and 10 mg of herbicide were dissolved in 10 mL of chloroform to form the organic phase. The aqueous phase (200 mL) was prepared using 0.5 % (w/v) polyvinyl alcohol, at 50 ºC. The organic phase was transferred to the aqueous phase (at 50 ºC) with magnetic stirring (1000 rpm for 15 min). The chloroform was then evaporated from the emulsion. The suspension of microparticles formed was stored in an amber flask (to avoid any photodegradation of the herbicide). The final concentration of ametryn was 50 mg L^{-1}.

Formulation	PHBV (mg)	%	PHB (mg)	%
A	200	100	0	0
B	150	75	50	25
C	100	50	100	50
D	50	25	150	75
E	0	0	200	100

Table 1. Proportions of polymers used to prepare the different formulations.

2.2.3 Measurements of encapsulation efficiency

Portions (10 mg) of the different microparticles containing herbicide were dissolved in 50 mL of acetonitrile, and the association rate of the herbicide with the microparticles was determined by the technique described previously, which involves ultrafiltration/ centrifugation and analysis using HPLC (Kilic et al., 2005; Schaffazick et al., 2003). The samples were centrifuged in regenerated cellulose ultrafiltration filters that had a molecular size-exclusion pore size of 30 KDa (Microcon, Millipore), and the filtrate was analyzed using HPLC. The ametryn concentration was obtained from an analytical curve. The association rate of ametryn was calculated from the difference between the concentration measured in the filtrate and the total concentration (100 %) in the microparticle suspension. The total concentration was measured after diluting the suspension with acetonitrile, which dissolved the polymer and ensured complete release of the herbicide. The measurements were performed in triplicate for each formulation. The encapsulation efficiency (EE, %) was expressed as the ratio:

$$EE(\%) = \frac{W_S}{W_{TOTAL}} \times 100\% \tag{1}$$

Where, W_s is the quantity of ametryn in the microparticles and W_{total} is the amount of ametryn used in the formulation.

2.2.4 Scanning electron microscopy (SEM)

A scanning electron microscope (Model JSM-6700F, JEOL, Japan) was used to investigate the size distribution and surface morphology of the microparticles. Suspensions of microparticles containing the herbicide were filtered and the particles were then washed with 150 mL of distilled water. The solid residues were dried overnight over Na_2SO_4 in a desiccator. The samples were then attached to metallic supports (stubs) with double-sided tape, and metalized by deposition of a gold layer at a current of 25 mA for 150 s. Images (electron micrographs) of the samples were then generated using the microscope. Particle sizes were measured using the ImageJ 1.42 program, and the size distributions of the different microparticles were obtained using OriginPro 7.0. At least 1000 individual particles of each sample were used for these measurements.

2.2.5 Release of ametryn from the microparticles

The release profiles of ametryn, either free or associated with the microparticles, were investigated using a two-compartment experimental system. A cellulose membrane (Spectrapore, with a molecular exclusion pore size of 1000 Da) separated the donor compartment, containing 4 mL of solution (or suspension) of the herbicide, from the acceptor compartment, which contained 50 mL of deionized water maintained under gentle agitation at ambient temperature (Paavola et al., 1995). The pore size of the membrane only allowed passage of the free herbicide, while the herbicide associated with the microparticles was retained in the donor compartment until the equilibrium was shifted so as to release the ametryn present within the particles. The size of the microparticles prevented their passage through the pores of the membrane. These experiments were conducted under *dilution sink* conditions, whereby the volume of the dissolution medium was sufficiently large that the herbicide concentration never exceeded 10 % of the value of its saturation concentration (Aulton et al., 2002).

Samples were retrieved from the acceptor compartment as a function of time, and analyzed by HPLC at a detector wavelength of 260 nm. During the first hour, samples were collected every 15 min, during the second hour every 30 min, and subsequently at hourly intervals until the peak area stabilized. The peak area values were then converted into the percentage of herbicide released as a function of time (De Araújo et al., 2004).

2.2.5.1 Mathematical modeling of ametryn release

Mathematical modeling is increasingly used to investigate the release profiles of bioactive compounds in polymeric systems, since it can provide important information concerning the release mechanism. Analysis of the mechanism of release of ametryn from the microparticles employed the zero order, first order, Higuchi and Korsmeyer-Peppas models (Colombo et al., 1995, 2005; Costa & Lobo, 2001; Ferrero et al., 2000; Hariharam et al., 1994; Ritger & Peppas, 1987a, 1987b).

3. Results and discussion

The encapsulation efficiency values obtained for the different microparticles are listed in Table 2. Formulation A (100 % PHBV) showed the highest encapsulation efficiency (76.5 %). The efficiency decreased as the proportion of PHBV decreased, and formulation E (100 % PHB) provided the lowest encapsulation efficiency (26.2 %). The values obtained for formulations A and B were fairly high, relative to values that have been reported in the literature for other active principles (Bazzo et al., 2009; Grillo et al., 2010; Lobo et al., 2011; Sendil et al., 1999). Grillo and colleagues (2010) showed that the encapsulation efficiency of the herbicide atrazine in PHBV microparticles was in excess of 30 %. Lobo et al. (2011), using an experimental design optimization procedure, obtained an encapsulation efficiency of 24 % for atrazine in PHBV microparticles.

Formulation	PHBV (%)	PHB (%)	EE (%)
A	100	0	76.5
B	75	25	54.7
C	50	50	40.5
D	25	75	29.3
E	0	100	26.2

Table 2. Encapsulation efficiencies (EE, %) of the different microparticles.

The relationship between the percentage of PHBV and the encapsulation efficiency is illustrated in Figure 2. There was a polynomial relationship between the encapsulation efficiency and the PHBV concentration, which was positive for PHBV and negative for PHB. This can probably be explained by the structural differences between the microparticles, due to the different polymer ratios used in their preparation (Table 1).

The morphological characteristics of the microparticles, as well as the influence of the encapsulation of ametryn, were analyzed using the SEM procedure. Electron micrographs of the microparticles containing ametryn are illustrated in Figure 3. All types of microparticle were spherical, although the surface structures were different. Most of the PHB microparticles possessed smooth surfaces with few pores, while most of the PHBV microparticles were rough-surfaced with many cavities and pores, some of which were quite large, as can be clearly seen for formulation A (Figure 3, a1 and a2). Grillo et al. (2010) also found that PHBV microparticles, prepared using the same methodology as that described here, were rough-surfaced with pores, while PHB microparticles had smooth surfaces and fewer pores.

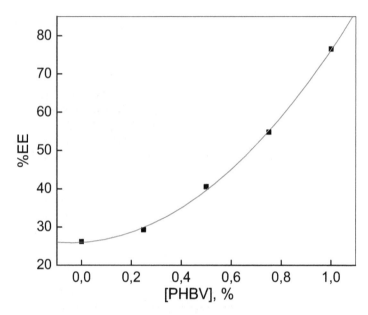

Fig. 2. Encapsulation efficiency according to PHBV content of the microparticles.

A higher encapsulation efficiency of ametryn was therefore related to a greater number of pores in the microparticles, probably due to greater contact (and/or affinity) of the herbicide with the microparticles during the formulation preparation procedure. Ametryn is likely to have greater affinity for the PHBV polymer, since both of these molecules possess alkyl branches, with interaction being further enhanced by the porosity of the PHBV microparticles.

The size distribution profiles (Figure 4) differed between microparticle formulations (it was not possible to measure the size distribution of the formulation D microparticles due to focusing problems). The average size of the microparticles (Table 3) increased as the PHBV concentration decreased and the PHB concentration increased, and was greatest for the PHB microparticles (formulation E). These size differences could be related to the incorporation of the herbicide as well as to associations between the molecules (as discussed above). At higher encapsulation rates, the amount of ametryn present within the microparticle increased, and the potential for reactions and interactions with the polymer therefore also increased. Ametryn is likely to have a higher affinity for PHBV, and as a result of this affinity (and/or reaction) the polymer contracts due to the formation of linkages between the polymer chains. As the proportion of PHBV decreases, the affinity of ametryn for the polymer mixture also diminishes (due to the lower affinity of ametryn for PHB), so that there is less shrinkage.

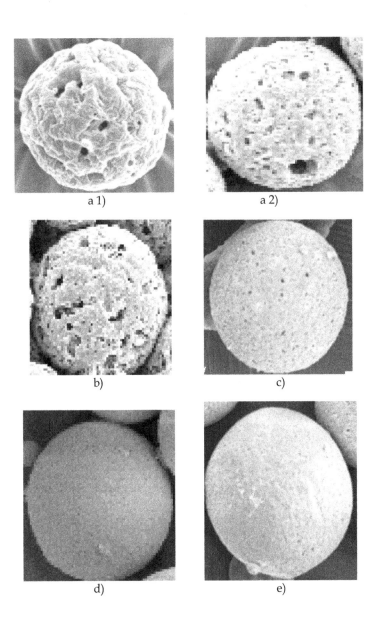

Fig. 3. SEM images of the polymeric microparticles: a) Formulation A; b) Formulation B; c) Formulation C; d) Formulation D; e) Formulation E.

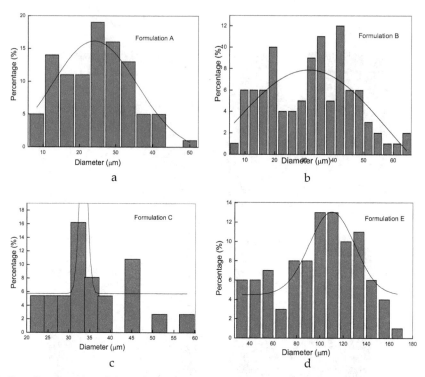

Fig. 4. Size distributions of the polymeric microparticles: a) Formulation A; b) Formulation B; c) Formulation C; d) Formulation E.

Formulation	PHBV (%)	PHB (%)	Average size (μm)
A	100	0	24.14 ± 1.606
B	75	25	31.45 ± 2.797
C	50	50	33.5 ± 3.22
D	25	75	*
E	0	100	110.2 ± 3.881

* Not determined.

Table 3. Average sizes (± SD) of the different microparticles.

The release profiles of free ametryn (as the reference) and ametryn encapsulated in the microparticles are illustrated in Figure 5, as a function of time (up to approximately 360 min). In these experiments the herbicide could traverse the pores of the membrane, while the microparticles were retained, so that it was possible to measure the influence of the association of ametryn with the polymeric matrix of the microparticles on its release rate. The release kinetics of free ametryn was faster than that of the encapsulated herbicide, with almost total release after 360 min. Association with the microparticles resulted in retarded release, with around 70 % (formulations A and B), 30 % (formulation C), 20 % (formulation D) and 40 % (formulation E) being released after 360 min.

The release of other bioactive compounds from systems composed of microstructured polymers has been described in the literature, but usually for only one type of polymer (Grillo et al., 2010; Maqueda et al., 2009; Sendil et al., 1999; Singh et al., 2010; Wang et al., 2007). However, interpretation of release profiles relies to a large extent on knowledge of the composition and structural characteristics of the microparticles concerned, and in this respect studies that use more than one type of microparticle are advantageous. In the present work, the release of ametryn increased in line with the content of PHBV for formulations A-D, indicating that increased porosity aided the exit of ametryn molecules due to increased contact with the solvent. However formulation E was an exception to the rule, since it was composed of PHB alone and showed the fastest release of ametryn. There are two possible explanations for this observation. Firstly, the encapsulation efficiency of this formulation was lower than those achieved using the other formulations, which could have resulted in higher concentrations of ametryn crystals in the solution, and consequently higher release rates. Secondly, it is possible that lengthy refrigerated storage of this sample could have resulted in solubilization of the herbicide, due to increased contact time with the solvent.

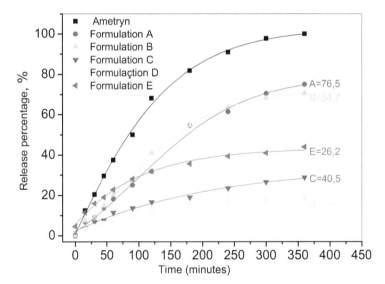

Fig. 5. Results of the release experiments, comparing the kinetic profiles of free ametryn and ametryn associated with the different microparticles (PHB, PHBV and PHBV+PHB), at ambient temperature (n = 3).

Analysis of release curves can provide important information concerning the mechanisms involved in the release of compounds from microparticles (Polakovic et al., 1999). Possible mechanisms include desorption from the surface of the polymeric matrix, diffusion through the pores or wall of the matrix, disintegration of the microparticle with subsequent release of the active principle, and dissolution and erosion of the matrix or the polymeric wall (Polakovic et al., 1999; Schaffazick et al., 2003).

A number of mathematical models have been extensively used to analyze the characteristics of the release of substances from polymeric systems (Costa & Lobo 2001). Here, the results of the release experiments (Figure 5) were analyzed using the zero order, first order, Higuchi and Korsmeyer-Peppas models (Table 4). For the formulations investigated, the Korsmeyer-Peppas model provided the best explanation of the ametryn release mechanism, according to the correlation coefficient obtained. The curves obtained for each formulation using this model are illustrated in Figure 6.

	Zero order	First order	Higuchi	Korsmeyer-Peppas
Formulation A				
				$n = 0{,}82641$
Release constant (k)	4.59184 min^{-1}	0.00667 min^{-1}	4.49925 $min^{-1/2}$	$0.00628 min^{-n}$
Correlation coefficient (r)	0.92307	0.98452	0.97721	0.99364
Formulation B				
				$n = 0.79373$
Release constant (k)	0.20767 min^{-1}	0.00624 min^{-1}	4.35701 $min^{-1/2}$	$0.0072 min^{-n}$
Correlation coefficient (r)	0.89455	0.96782	0.98115	0.9879
Formulation C				
				$n = 0.62532$
Release constant (k)	0.07283 min^{-1}	0.00581 min^{-1}	1.52624 $min^{-1/2}$	0.0162 min^{-n}
Correlation coefficient (r)	0.86545	0.97701	0.9893	0.9929
Formulation D				
				$n = 0.5671$
Release constant (k)	0.048 min^{-1}	0.00495 min^{-1}	1.00913 $min^{-1/2}$	$0.0194 min^{-n}$
Correlation coefficient (r)	0.90337	0.96059	0.97587	0.98839
Formulation E				
				$n = 0.42726$
Release constant (k)	0.0983 min^{-1}	0.00441 min^{-1}	2.17047 $min^{-1/2}$	$0.0429 min^{-n}$
Correlation coefficient (r)	0.79093	0.92828	0.99035	0.9927

Table 4. Results of the application of four mathematical models to the release curves of ametryn associated with different microparticles.

The Korsmeyer-Peppas model is based on a semi-empirical equation (Korsmeyer & Peppas, 1991; Korsmeyer et al., 1983) that is widely used when the release mechanism is unknown. When the release exponent (n) is equal to 0.43 the mechanism involved is diffusion. When the value of the exponent is greater than 0.43 but smaller than 0.85, the release occurs due to anomalous transport that does not obey Fick's Law. Values less than 0.43 are indicative of porous systems in which transport occurs by a combination of diffusion through the polymeric matrix and diffusion through the pores. The values obtained (Table 4) differed according to formulation, as expected considering the different structural characteristics of the microparticles, so that the release mechanisms were not identical. Nonetheless, the values obtained for all formulations were in the range $0.43 < n < 0.85$, indicating that in all cases the release occurred as a result of anomalous transport, involving diffusion and relaxation of the polymeric chains. This information concerning the release mechanism is of vital importance in order to be able to adjust and optimize the release of the active principle according to circumstances.

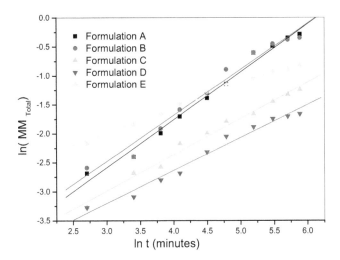

Fig. 6. Results obtained using the Korsmeyer-Peppas model applied to formulations A-E.

4. Conclusions

Ametryn herbicide was efficiently encapsulated in microparticles composed of PHB, PHBV and mixtures of the two polymers. The highest encapsulation efficiencies were achieved when higher proportions of PHBV were used. SEM analysis showed that the microparticles were spherical, although with different surface features (either smooth or rough with pores). The release profile of ametryn was modified when it was encapsulated, with slower and more sustained release compared to the free herbicide. This finding suggests that the use of encapsulated ametryn could help to mitigate adverse impacts on ecosystems and human health. This is particularly important given the increasingly widespread and intensive use of agents such as ametryn in modern agriculture.

5. Acknowledgments

The authors thank FAPESP, CNPq and Fundunesp for financial support.

6. References

Ahmadi, A. & Ahmadi, A. (2007). Preparation and characterization of chemical structure composition of polyurethane's microcapsules pesticides. *Asian Journal of Chemistry*, Vol.19, No.1, pp. 187-194.

Amass, W. & Tighe, B. (1998). A review of biodegradable polymers: uses, current developments in the synthesis and characterization of biodegradable polyesters, blends of biodegradable polymer and recent advances in biodegradation studies. *Polym. Int.*, Vol.47, pp. 89-144.

Aulton, M. (2002). Dissolution and Solubility. Pharmaceutics: the science of dosage from design. 2 ed. Edinburgh: Churchill Livingstone, cap. 2, p. 15-32,

Bazzo, G.C.; Lemos-Senna, E. & Pires, A.T.N. (2009). Poly(3-hydroxybutyrate)/chitosan/ ketoprofen or piroxicam composite microparticles: Preparation and controlled drug release evaluation, *Carbohydr. Polym.*, Vol.77, pp. 839–844.

Bin Hussein, M.Z.; Hashim, N.; Yahaya, A.H. & Zainal, Z. (2009). Controlled Release Formulation of Agrochemical Pesticide Based on 4-(2,4-dichlorophenoxy)butyrate Nanohybrid. *Journal of nanoscience and nanotechnology*, Vol.9, No. 3, pp. 2140-2147.

Coimbra, P.A.; De Sousa, H.C. & Gil, M.H. (2008). Preparation and characterization of flurbiprofen-loaded poly(3-hydroxybutyrate-co-3- hydroxyvalerate) microspheres. *J Microencapsul* ., Vol. 25; pp.170-178.

Colombo, G.; Padera, R.; Langer, R. & Kohane, D.S. (2005). Prolonged duration anesthesia with lipid-protein-sugar particles containing bupivacaine and dexamethasone. *J Biomed Mater Res Part A* , Vol.75; pp. 458-464.

Colombo, P.; Bettini, R.; Massimo, G.; Catellani, P.L.; Santi, P. & Peppas, N.A. (1995). Drug diffusion front movement is important in drug release control from swellable matrix tablets. *J Biomed Mater Res.*, Vol.84, pp. 991-997.

Conti, B.; Genta, I.; Modena, T. & Pavanetto, F. (1995). Testing of In Vitro Dissolution Behaviour of Microparticulate Drug Delivery Systems. *Drug Dev. Ind. Pharm.*, Vol.21, pp. 1223-1233.

Costa, P. & Lobo, J.M.S. (2001). Modeling and comparison of dissolution profiles. *Eur J Pharm Sci*, Vol.13, pp. 123-133.

De Araújo, D.R.; Cereda, C.M.S.; Brunetto, G.B.; Pinto, L.M.A.; Santana, M.H.A. & De Paula, E. (2004). Encapsulation of mepivacaine prolongs the analgesia provided by sciatic nerve blockade in mice. *Can. J. Anesth.*, Vol.51, pp. 566-572.

El Bahri, Z. & Tavedert, J.L. (2007). Elaboration and characterisation of microparticles loaded by pesticide model. *Powder Technol.*, Vol.172, pp. 30-40.

El Bahri, Z. & Taverdet, J.L. (2005). Optimization of an herbicide release from ethylcellulose Microspheres. *Polymer Bulletin.*, Vol.54, pp. 97-103.

Ferrero, C.; Muñoz-Ruiz, A. & Jiménez-Castellanos, M.R. (2000) Fronts movement as a useful tool for hydrophilic matrix release mechanism elucidation. *Int J Pharm.*, Vol.202, pp. 21-28.

Friedmann, A.S. (2002) *Atrazine inhibition of testosterone production in rat males following peripubertal exposure.* Johns Hopkins University, Baltimore, MD, USA.

Grillo, R.; Melo, N.F.S.; Lima, R.; Lourenço, R.; Rosa, A.H. & Fraceto, L.F. (2010). Characterization of atrazine-loaded biodegradable poly(hydroxybutyrate-co-hydroxyvalerate) microspheres. *J Polym Environ.*, Vol.18, pp. 26-32.

Grillo, R.; Melo, N.F.; Moraes, C.M.; De Lima, R.; Menezes, C.M.; Ferreira, E.I.; Rosa, A.H. & Fraceto, L.F. (2008). Study of the interaction between hydroxymethylnitrofurazone and 2-hydroxypropyl-beta-cyclodextrin. *Pharm Biomed Anal.*, Vol. 47, pp. 295-302.

Hariharam, D.; Peppas, N.A.; Bettini, R. & Colombo, P. (1994) Mathematical analysis of drug delivery from swellable systems with parcial physical restrictions or impermeable coatings. *Int J Pharm.*, Vol.112; pp. 47-54.

Hirech, K,; Payan, S.; Carnelle, G.; Brujes, L. & Legrand, J. (2003). Microencapsulation of an insecticide by interfacial polymerization. *Powder Technol.*, Vol.130, pp. 324-330.

Kilic, A.C.; Capan, Y.; Vural, I.; Gursoy, R.N.; Dalkara, T.; Cuine, A. & Hincal, A.A. J. (2005). Preparation and characterization of PLGA nanospheres for the targeted delivery of NR2B-specific antisense oligonucleotides to the NMDA receptors in the brain. *Microencap*, Vol. 22, pp. 633–641.

Korsmeyer, R.W. & Peppas, N.A. Macromolecular and modeling aspects of swelling-controlled systems. In: Roseman TJ, Mansdorf SZ (Eds.). Controlled release delivery systems. New York, USA: Marcel Dekker Inc, 1991.

Korsmeyer, R.W.; Gurny, R.; Doelker, E.; Buri, P. & Peppas, N.A. (1983). Mechanisms of Solute Release from Porous Hydrophilic Polymers, *Int. J. Pharm.*, Vol.15, pp. 25-35.

Li, J.Y.; Zu, B.Y.; Zhang, Y.; Guo, X.Z. & Zhang, H.Q. One-Pot Synthesis of Surface-Functionalized Molecularly Imprinted Polymer Microspheres by Iniferter-Induced "Living" Radical Precipitation Polymerization *Journal Of Polymer Science Part A-Polymer Chemistry*, Vol.48, pp. 3217-3228.

Lionzo, M.I.Z.; Ré, M.I., Guterres, S.S. & Pohlmann, A.R. J. (2007). Microparticles prepared with poly(hydroxybutyrate-co-hydroxyvalerate) and poly(ε-caprolactone) blends to control the release of a drug model. *Microencapsulation*, Vol.24, pp. 175-186.

Lobo, F.A.; Aguirre C.L.; Silva, M.S.; Grillo, R.; Melo, N.F.S.; Oliveira, L.K.; Morais, L.C.; Campos, V.; Rosa, A.H. & Fraceto, L.F.; (2011). Poly(hydroxybutyrate-co-hydroxyvalerate) microspheres loaded with atrazine herbicide: screening of conditions for preparation, physico-chemical characterization and *in vitro* release studies. *Polym. Bull.* Vol.67, pp. 479–495.

Maqueda, C.; Villaverde, J.; Sopeña, F.; Undabevtia, T. & Morillo, E. (2009). Effects of soil characteristics on metribuzin dissipation using clay-gel-based formulations. *J. Agric. Food Chem*, Vol.57, pp. 3273-3278.

Melo, N.F.; Grillo, R.; Rosa, A.H&; Fraceto L.F.J (2008). Interaction between nitroheterocyclic compounds with β-cyclodextrins: Phase solubility and HPLC studies. *Pharm Biomed Anal.*, Vol.47, No. 4-5, pp. 865-869.

Moraes, C.M.; De Paula, E.; Rosa, A.H. & Fraceto, L.F. Physicochemical stability of poly(lactide-co-glycolide) nanocapsules containing the local anesthetic Bupivacaine. *Journal of the Brazilian Chemical Society*, v.21, p. 995-1000, 2010.

Natarajan, V.; Krithica, N.; Madhan, B. & Sehgal, P.K. (2011). Formulation and evaluation of quercetin polycaprolactone microspheres for the treatment of rheumatoidaArthritis. *Journal of pharmaceutical sciences*, Vol.100, No.1, pp.195-205.

Paavola, A.; Yliruusi, J.; Kajimoto, Y.; Kalso, E.; Wahlström, T. & Rosenberg, P. (1995). Controlled release of lidocaine from injectable gels and efficacy in rat sciatic nerve block. *Pharm. Res.*, Vol.12, pp. 1997-2002.

Parajo, Y.; d'Angelo, I.; Horvath, A.; Vantus, T.; Gyorgy, K.; Welle, A.; Garcia-Fuentes, M. & Alonso, M.J. (2010). PLGA:poloxamer blend micro- and nanoparticles as controlled release systems for synthetic proangiogenic factors. *European Journal of Pharmaceutical Sciences*, Vol.41 No.5, pp.644-649.

Polakovic, M.; Gorner, T.; Gref, R. & Dellacherie, E. (1999) Lidocaine loaded biodegradable nanospheres II: Modelling of drug release. *J. Controlled Release*, Vol.60, pp. 169-177.

Pouton, C.W. & Akhtarb S. (1996). Biosynthetic polyhydroxyalkanoates and their potential in drug delivery. *Advanced Drug Delivery Reviews*, Vol. 18, pp. 133-162.

Reis, C.P.; Neufeld, R.J.; Ribeiro, A.J. & Veiga, F. (2006). Nanoencapsulation I: Methods preparation of drug-loaded polymeric nanoparticles. *Nanomedicine*, Vol.2, pp. 8-21.

Ritger, P.L. & Peppas, N.A. J. (1987a). A simple equation for description of solute release II. Fickian and anomalous release from swellable devices. *Control. Release*, Vol.5, pp. 37–42.

Ritger, P.L. & Peppas, N.A. J. (1987b). A simple equation for description of solute release I. Fickian and non-fickian release from non-swellable devices in the form of slabs, spheres, cylinders or discs. *Control. Release*, Vol.5, pp. 23–36;

Salehizadeh, H. & Van Loosdrecht, M.C.M. (2004). Production of polyhydroxyalkanoates by mixed culture: recent trends and biotechnological importance. *Biotechnol. Adv.*, Vol.22, pp. 261-279.

Schaffazick, S. R.; Guterres S. S.; Freitas, L. L. & Pohlmann, A. R. (2003). Caracterização e estabilidade físico-química de sistemas poliméricos nanoparticulados para administração de fármacos *Quim. Nova*, Vol.26, pp. 726-737.

Sendil, D.; Gursel. I.; Wise, D.L. & Hasircia, V. (1999) Antibiotic release from biodegradable PHBV microparticles, *J. Controlled Release*, Vol.59, pp. 207– 217.

Silva, M.D.; Cocenza, D.S.; de Melo, N.F.S.; Grillo, R.; Rosa, A.H. & Fraceto, L.F. (2010). Alginate nanoparticles as a controlled release system for clomazone herbicide. *Quimica Nova*, Vol.33, pp. 1868-1873.

Singh, B.; Sharma, D.K. & Gupta, A. J. (2008). In vitro release dynamics of thiram fungicide from starch and poly(methacrylic acid)-based hydrogels. *Hazardous Materials*, Vol.154, pp.278-286.

Singh, B.; Sharma, D.K.; Kumar, R. & Gupta, A. J. (2010). Development of a new controlled pesticide delivery system based on neem leaf powder. *Hazardous Materials*, Vol.177, pp. 290-299.

Sinha, V.R.; Bansal, K.; Kaushik, R.; Kumria, R. & Trehan, A. (2004). Poly-ε-caprolactone microspheres and nanospheres: an overview. *Int. J. Pharm.*, Vol.278, pp. 1–23.

Sopeña F.; Maqueda, C. & Morillo E. (2009). Controlled release formulations of herbicides based on micro- encapsulation. *Cien. Inv. Agr.* Vol.35, No.1, pp. 27-42.

Sudesh, K.; Abe H.& Y. Doi. (2000). Synthesis, structure and properties of polyhydroxyalkanoates: biological polyesters. *Prog. Polym. Sci.* Vol.25, pp.1503-1555.

Tennant, A.H.; Peng, P. & Kligerman, A.D. (2001). Genotoxicity studies of three triazine herbicides: in vivo studies using the alkaline single cell gel (SCG) assay. *Mutation Research*, Vol.493, No.1-2, pp. 1–10.

Wang, Y.; Wang, X.; Wei, K.; Zhao, N.; Zhang, S. & Chen, (2007). Fabrication, characterization and long-term in vitro release of hydrophilic drug using PHBV/HA composite microspheres. *J. Materials Letters*, Vol.61, pp. 1071–1076.

4

The Influence of Biochar Production on Herbicide Sorption Characteristics

S.A. Clay and D.D. Malo
South Dakota State University, Plant Science Dept.
Brookings, South Dakota
USA

1. Introduction

Biochar is the by-product of a thermal process conducted under low oxygen or oxygen-free conditions (pyrolysis) to convert vegetative biomass to biofuel (Jha et al., 2010). There are a wide variety of end-products that can be manufactured depending on processing parameters and initial feedstocks (Bridgewater, 2003). The pyrolytic process parameters such as temperature, heating rate, and pressure can change the recovery amounts of each end-product, energy values of the bio-oils, and the physico-chemical properties of biochar (Yaman, 2004).

Biochars are recalcitrant forms of carbon and, depending on properties, can remain in the soil for greater than 1000 years (Skjemstad et al., 2002). The long-term persistence of this carbon form is due to slow microbial degradation and chemical oxidation rates (Sanchez et al., 2009). In addition, biochar interacts with soil materials such as ions, organic matter, and clays that generally increase the persistence of biochar within the soil. However, biochars, unlike commercial fertilizers, are not precisely defined materials and vary widely in properties depending on organic material source and manufacturing process (Karaosmanoglu et al., 2000; McHenry, 2009; Sohi et al., 2010). Increasing pyrolytic temperature decreases biochar recovery but increases C concentration of the char compared with char recovered at lower temperatures (Daud et al., 2001; Katyal et al., 2003). For example, as temperature increased from 300^0 to 800^0 C, biochar C content increased from 56 to 93% whereas biochar yield decreased from 67 to 26% (Okimori et al., 2003). Other pyrolytic parameters, such as sweep gas flow, can influence biochar particle size with higher flows reducing the particle size but increasing heating values (Katyal et al., 2003; Demirbas, 2004). Biochar also can be influenced by reactor design and other reaction parameters including heating rate, residence time, pressure, and catalyst used. Feedstock type, quality, and initial physical characteristics of the material (e.g. particle size, shape, and structure) can impact the bio-oil yield and properties, as well as the type and amounts of biochar formed (Bridgewater et al., 1999).

Landspreading biochar for a soil amendment is suggested to improve crop production efficiency because regardless of the initial manufacturing process, biochars have a high charge density and surface area. The use of biochar as a soil amendment is not a new concept. Dark earths (terra preta) discovered in the Amazon Basin were found to have received deliberate land applications of charred materials and residues of biomass burning by Amer-indian populations before European arrival (Erickson, 2003; Sombroek et al. 2003). Pyrogenic C in terra preta is very resistant to microbial decay over centuries due to its complex aromatic structure and acts as a significant C sink (Glaser et al., 2001).

The benefits of biochar application have been hypothesized to include: increasing plant available soil water; building soil organic matter; enhancing nutrient cycling; lowering soil bulk density; acting as a liming agent if high in pH; and reducing transfer of pesticides and nutrients to surface and ground water (Laird, 2008) thereby improving water quality. The application of biochar to soil has been reported to have a positive impact on physical properties such as soil water retention and aggregation (Piccolo et al., 1996) and may decrease erosion potential. Glaser et al. (2002) observed an increase in field water holding capacity by 18% in charcoal enriched Anthrosol due to an increase in surface area. Biochar application has been shown to improve other soil physical, chemical, and biological properties (Glaser et al., 2002; Lehmann and Rondon, 2006) leading to positive impacts on plant growth and development. For example, Chidumayo (1994) observed enhanced seed germination (30%), shoot height (24%), and biomass production (13%) of seven indigenous woody crops with the application of charcoal compared with the crops on undisturbed Zambian Alfisols and Ultisols. Kishimoto and Sugiura (1985) also found increases in height (26 to 35%) and biomass (2.3 X greater) production of sugi trees (*Cryptomeria japonica* L.). Similar enhancement was observed in yields of annual crops such as maize (*Zea mays* L.) on Nigerian Alfisols and Inceptisols with the application of charcoal (Mbagwu and Piccolo, 1997) due to an increase of soil pH that resulted in greater micro-nutrient availability and decreased deficiencies. However, biochars also have been shown to have an extreme affinity for essential plant   nutrients (Sanchez et al., 2009) that can provide a slow release mechanism.

Some biochars that have high pH (e.g. >9.5) can provide liming capacity and increase the soil pH (Sanchez et al., 1983; Mbagwu and Piccolo, 1997). For example, application of coal ash at the rate of 110 Mg ha[-1] increased the pH of an eroded Palouse soil from 6.0 to 6.8 (Cox et al., 2001). Exchangeable bases also were observed to increase in sandy and loamy soils with the additions of hardwood and conifer charcoals (Tryon, 1948). Application of charcoal to highly weathered soils with low-ion retention capacities increased the cation exchange capacity (CEC) by 50% compared to unamended soil (Mbagwu and Piccolo, 1997). Oguntunde et al. (2004) reported a significant increase in soil pH, base saturation, electrical conductivity (EC), exchangeable Ca, Mg, K, Na, and available P in charcoal kiln sites and reported an increase in grain and biomass yield of maize of 91% and 44% respectively, with a coal char application. Leaching of NH_4^+ from an unfertilized Ferralsol was reduced with the application of charcoal due to its high C content, although the retention properties of chars may differ for other ionic species (e.g. K, Ca, Mg) if the char already contains high concentrations of the ion of interest (Lehmann et al., 2002). Because of biochar's diverse properties and potential for high reactivity in soils, a 'one-recommendation-fits-all situations' mentality for the use as of biochar as a soil amendment needs to be avoided. To date, the greatest positive impacts of biochar have been primarily observed on degraded soils and those with low fertility whereas applications on highly productive soils have been reported to have low or minimal impacts (Woolf et al., 2010).

Agrichemicals such as pesticides, growth regulating chemicals, and nutrients are applied to crops to control pests and increase yield potential. Depending on the type and amount of biochar applied, the changes in soil properties associated with the application (e.g. soil pH, EC) as well as the physio-chemical properties of the char itself, may impact the use, rates, efficacious properties, and fates of agrichemicals used in agronomic management. The environmental fate (e.g. leachability, rate of decomposition, etc.) and efficacy of soil applied pesticides are influenced strongly by their reaction and retention with soil particles and organic matter (Brown et al., 1995). Agrichemical molecules can be removed from soil solution through attraction or attachment to the surfaces of organic materials and soil particles (adsorption) or movement into the matrix (like water into a sponge) (absorption). Often, experiments cannot distinguish between these processes so that the general term sorption is used.

Sorption is controlled by properties of the chemical of interest including the water solubility, pH, dissociation constant (pKa), octanol/water partition coefficient, and other factors (Weber, 1995) and can be used to help describe the fate of an herbicide in the environment (Wauchope et al., 2002). The sorption of the chemical also is affected by soil properties including water, organic matter, clay, sand, and oxide contents, and soil pH (Koskinen and Clay, 1997; Laird and Koskinen, 2008). Soils high in sand generally sorb much less chemical than loamy or clay type soils. Agricultural practices that involve modifying soil organic matter content often increase chemical retention. Indeed, studies have shown that adding biochar to soil can result in greater sorption of pesticides (Cao et al., 2009; Spokas et al., 2009; Yu et al., 2009). The distribution of chemical between a solution and solid phase gives an indication of the amount of chemical available in solution and is defined using a sorption coefficient (K_d) where:

$$K_d = \frac{\text{mass of herbicide sorbed per g of solid}}{\text{amount of chemical remaining in solution at equilibrium}} \qquad (1)$$

Large K_d values (typically over 100) indicate that a high amount of the chemical originally in solution is sorbed to the solid interface, with low amounts of chemical remaining in solution. Sorption of a chemical from the liquid phase of soil may result in the chemical being: 1) less available to plants, so there may be less uptake; 2) less available to soil organisms, thereby increasing the chemical's residence time and slowing degradation; and 3) less available to leach with water percolating through the soil, which could result in improved groundwater quality.

The biochar source-processing combination provides a rich diversity of biochars to evaluate for soil amendment use (Lehmann et al., 2009). The potential of a specific biochar for a specific use will depend on the physical and chemical properties of the biochar, as well as soil characteristics. The challenge of amending soil with biochar is to identify the benefits that biochar can provide (e.g. fertility, increased water holding capacity) (Lehmann, 2007) and balance these against any negative effects that the char may have. Site-specific application recommendations of specific biochars require an examination of the products of different production and processing scenarios. Much of the biochar research has been based on slow pyrolysis with a goal to optimize biochar properties for a specific goal such as improved soil fertility, greenhouse gas mitigation, or heating value. Little work has been done with biochar produced from fast pyrolysis processes and even less with biochar produced from microwave pyrolysis reactors.

Feedstock is a key factor governing the status of physio-chemical properties of biochar. All types of materials including, but not limited to, palm shells, rapeseed (*Brassica rapa*) stems, sunflower (*Helianthus annuus*), and wood have been used or are being proposed as potential feedstock sources for use in the biofuel industry. In the Midwestern U.S., maize stover and switchgrass (*Panicum virgatum*) biomass are feedstocks that bioenergy companies are exploring for use.

2. Biochar influence on herbicide sorption to soil

This study examined atrazine and 2,4-D sorption to several biochars that were the result of microwave pyrolysis using varying temperatures and processing times of maize and switchgrass biomass. In addition, sorption characteristics of these two chemicals to soil amended with these biochars at two application rates were determined.

2.1 Materials and methods

2.1.1 Biochar and soil

Biochar was produced from maize stover (stalks and other residues remaining after maize grain harvest) and switchgrass biomass collected from fields near Brookings, South Dakota, USA (44.31, -96.67). Briefly, the material was dried at room temperature and pulverized mechanically using a Thomas-Wiley laboratory mill (Model No. 3375-E15, Thomas Scientific, USA) to pass through a 4 mm screen. The ground materials were processed by microwave pyrolysis using the SDSU Ag and Biosystem Eng. Dept. microwave system (specific processing methods reported in Lei et al., 2009). Processing temperatures ranged from 530^0 to 670^0 C and microwave residence times ranged from 8 to 24 minutes with seven maize and nine switchgrass biochars produced (Table 1 and Figures 1 and 2). The energy output, product types, particle size distribution, and elemental analysis of the biochar recovered from maize stover using these processing conditions are reported in Lei et al. (2009).

For this study, the maize biochars were used alone or mixed with the A horizon soil of a Brandt silty clay loam (Fine-silty, mixed, superactive, frigid Calcic Hapludoll, [Soil Survey Staff, 2011]) soil at 1 or 10% (w/w) to examine their effect on solution pH, EC, and atrazine and 2,4-D K_ds (sorption coefficients) for each biochar and biochar/soil combination. For switchgrass biochars, the 1 or 10% amendments to soil were used for pH and EC measurements, however, for herbicide sorption studies only biochar alone or soil mixed with 10% biochar were used, due to limited biochar supply. To maximize homogeneity, each soil/biochar combination was individually mixed by adding air-dry soil and biochar to each individual tube.

2.1.2 Solution characteristics

Biochars, soil, and soil with biochar amendments were analyzed for pH using a 0.01 M $CaCl_2$ slurry (1:1 w/v) and a standardized pH electrode. The solution pH was recorded after the reading had stabilized. Electrical conductivity (EC) was determined on a slurry that was mixed 1:1 (v/w) with 0.01 M $CaCl_2$ and biochar, soil, or soil amended with biochar. The slurry was shaken for 0.5 hr and EC measured using a commercially available EC electrode.

2.1.3 Herbicide sorption

Atrazine solution was diluted to a final concentration of 13 µM in 0.01 $CaCl_2$ using technical grade atrazine. This solution was spiked with about 0.4 kBq of uniformly-ring-labeled [^{14}C] atrazine (specific activity of 1000 MBq mmol^{-1} with > 99% purity; Sigma Chemical Co., St. Louis, MO). The 2,4-D solution was made in a similar manner, with technical grade 2,4-D added to 0.01 M $CaCl_2$ to have a final concentration of 13 µM. This solution was spiked with uniformly-ring-labeled [^{14}C]-2,4-D (specific activity of 1000 MBq mmol^{-1} with > 99% purity; Sigma Chemical Co., St. Louis, MO).

A 4-mL aliquot of herbicide solution was added to 2 g soil or soil amended with 1 or 10% biochar (final slurry solution 2:1 v/w) in glass centrifuge tubes sealed with a Teflon-lined cap. A 5-mL aliquot of herbicide solution was added to 0.5 g biochar when biochar was used as the sorbent, with the final solution/biochar ratio of was 10:1 v/w, due to the highly sorbent characteristics of the biochar.

After solution addition, the mixtures were shaken or vortexed to form a slurry. Tubes containing the slurries were shaken for 24 hr, centrifuged, and a 250-µL aliquot of supernatant removed. The amount of ^{14}C remaining in the supernatant solution was determined by liquid scintillation (Packard Model 1600TR) counting after the addition of scintillation cocktail. The amount of radioactivity sorbed was determined by comparing the counts in the supernatant samples with counts recorded from the original soil-free blank solution samples. The sorption coefficients (Kd) of the samples were then calculated as L kg^{-1}, correcting for the differences in volume added g^{-1} of material.

2.1.4 Statistical analysis

Experimental treatments were run in triplicate and studies were repeated in time. Results were combined for the studies due to similarity of means and homogeneity of variance between studies. Means presented were averaged over all treatment replicates and statistically separated by least significant difference calculation at $P \leq 0.05$.

2.2 Results

2.2.1 Biochar pH and EC values

The biochars produced in this study ranged in pH from acidic (4.06) to alkaline (9.88), and were dependent on feedstock, pyrolysis temperatures, and processing times (Table 1). Differences were observed among maize and switchgrass feedstocks. For maize stover, three of the microwave pyrolysis reactions at high temperatures ($\geq 650°C$), regardless of processing time, resulted in biochars that were very alkaline (pH>9). Two processes at lower temperatures (530°C and a processing time of 16 min or 550°C with a processing time of 10 min) resulted in biochars with pH <5. The 22 min processing time at 550°C resulted in a biochar with a more neutral (7.6) pH. For switchgrass, four processes resulted in biochars that were acidic (pH < 4.6) and the biochars were more acidic than biochars from maize at the same time and temperature. The acidic biochars were formed from processes that had low temperatures (<600°C) or shorter times at 600°C (8 min), or 10 min at 650ºC. The most alkaline switchgrass biochar was the result of processing at 670°C for 16 min. This biochar had a pH of ~9.1, which was lower than the alkaline maize biochars that ranged in pH from

Pyrolysis parameters		Maize (*Zea mays*)					
		pH			EC		
Temp	time	Biochar	soil + 1% biochar	soil + 10% biochar	Biochar	soil + 1% biochar	soil + 10% biochar
°C	min				mS cm⁻¹		
530	16	4.59	6.39	5.85⁻	0.3	2.4	1.4
550	10	4.77	6.38	6.04⁻	2.3	1.8	2.2
	22	7.60	6.47	6.61⁺	1.9	1.8	1.8
600	8	5.68	6.44	6.44	2.1	1.8	2.0
650	10	9.88	6.46	6.75⁺	2.0	1.9	1.9
	22	9.43	6.43	6.76⁺	2.0	1.8	1.9
670	16	9.65	6.43	6.73⁺	1.1	1.9	1.9
		Switchgrass (*Panicum virgatum*)					
530	16	5.32	6.17	6.70	0.3	1.80	1.67
550	10	4.12	6.49	5.67⁻	2.1	2.13	1.97
	22	4.06	6.49	5.71⁻	1.5	1.87	2.13
600	8	4.15	6.60	5.90⁻	1.9	1.80	1.83
	16	6.47	6.45	6.76⁺	1.7	1.30	1.33
	24	5.60	6.61	6.44	1.8	1.67	1.87
650	10	4.57	6.44	6.11⁻	2.0	2.07	2.20
	22	8.28	6.48	6.80⁺	2.9	1.97	2.37⁺
670	16	9.10	6.48	6.85⁺	2.5	1.67	1.90

Table 1. The influence of seven maize stover and nine switchgrass biochars produced with microwave pyrolosis with different processing times and temperature conditions on 100% biochar and soils amended with 1% or 10% (w/w) biochar. The soil used for this study was the A horizon of a Brandt silty clay loam (Fine-silty, mixed, superactive, frigid Calcic Hapludoll, [Soil Survey Staff, 2011]) from Aurora, SD (44.31, -96.67) with an unamended pH in a 1:1 solution of 0.01 M CaCl₂ of about 6.40 and an EC value of 1.63 mS cm⁻¹. A '-' sign indicates significantly lower value and a '+' sign indicates significantly higher value compared with unamended soil.

~9.4 to 9.9. The pH of these biochars can be compared with other biochar data. A wood ash/biochar that was the by-product of a commercial ethanol plant (Chippewa Valley Ethanol Company, Benson, MN) was obtained and used for comparison purposes. The wood ash had a pH of over 11. In comparison, broiler litter biochar obtained from pyrolysis reactions at either 350 or 700°C was found to have a fairly uniform acidic pH (5.5) (Uchimiya et al., 2010). These data indicate that the pH of different types of biochar are dependent on processing time, temperature, and initial feedstock material.

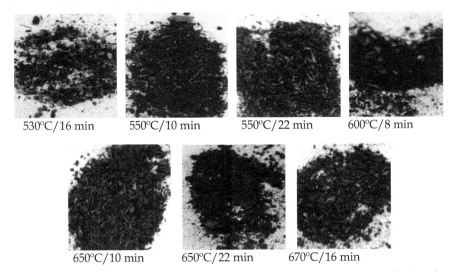

Fig. 1. Examples of biochars formed after exposure of maize (*Zea mays*) stover feedstocks to microwave pyrolosis at varying temperatures and times (see Lei et al., 2009).

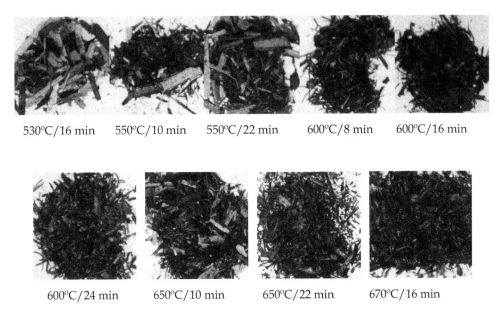

Fig. 2. Examples of biochars formed after exposure of switchgrass (*Panicum virgatum*) feedstocks to microwave pyrolysis at varying temperatures and times.

Electrical conductivity provides an indication of the amount of neutral soluble salts in the material or its salinity. High soil salinity often impedes the growth of most agricultural plants. Adding amendments that increase soil salinity, even though other beneficial properties such as water holding capacity would increase, would be counterproductive. Saline soils are recognized worldwide (Food and Agriculture Organization, FAO) as soils with an EC reading of >4 mS cm^{-1} (Richards, 1954; Abrol et al., 1988). In the U.S., the Soil Science Society of America (SSSA) uses a value of >2 mS cm^{-1} boundary for the saline classification. Woodchip biochar had an EC value of 3.6 mS cm^{-1}. Biochar produced from maize stover had EC values ranging from 1.1 to 2.3 mS cm^{-1} with five out of the seven >1.9 mS cm^{-1}. The switchgrass biochars had EC values ranging from 1.5 to 2.9 mS cm^{-1} with the highest EC when materials were processed at 650° C for 22 min.

2.2.2 Influence on biochar amendment on soil pH and EC properties

The Brandt soil chosen for this study was a silty clay loam with a pH of 6.4. Due to the inherent soil properties and buffering capacity of this soil, it was expected that even high applications of the most acidic or alkaline biochar would have minimal impact on soil pH. When 1% maize or switchgrass biochars were added to soil, pH changes were minimal (generally <3%) (Table 1). When soils were amended with 10% biochar, pH was influenced to a greater extent. The slurry pH decreased from 4 to 8% when low pH biochars were added and increased a maximum of 9% when high pH biochars were added.

Soil EC was 1.63 mS cm^{-1}, well below the salinity values for saline soil. Adding either maize or switchgrass biochar to soil at 1% increased soil salinity, but with the exception of one switchgrass sample, did not increase the salinity to >2 mS cm^{-1}. Amending soil with 10% with the maize biochar that had the greatest EC value (2.3 mS cm^{-1}) was the only maize biochar that increased soil salinity above 2 mS cm^{-1}. Adding switchgrass biochar at 10% had greater impact than maize stover biochar and increased EC values an average of 11% when compared with ECs of unamended soil. Three switchgrass biochars increased EC values from 23 to 36% (Table 1) with final soil slurry EC values above 2 mS cm^{-1}, the SSSA value for saline soil classification. However, even with a 10% amendment, all final EC values were well below the FAO saline soil value of 4 mS cm^{-1}. If significant amounts of these biochars were applied frequently to the same field, managers must be cognizant of the potential for changes to EC values. Saline soil remediation can be expensive and often requires long-term management interventions, rather than short-term programs.

2.2.3 Atrazine sorption to biochar and soils amended with biochar

Atrazine is a chemical in the triazine family and has a slightly positive charge in soil solutions (Laird and Koskinen, 2008). The positive charge on the molecule, when in solutions above its pK$_a$, causes the molecule to be sorbed to materials that have a negative charge. Atrazine sorption to soil is considered moderate with K$_d$ values ranging from 1 to 5 (Koskinen and Clay, 1997). The value is dependent on many soil properties including pH, organic matter, and clay content (Koskinen and Clay, 1997). In this study, atrazine sorption to biochar ranged from 7 to 92 L kg^{-1} (Figure 3). The sorption was dependent on feedstock type and processing method. These values ranged from 200 to 2300% greater than sorption to soil.

In general, the biochars from maize had much more variability in K$_d$ values than switchgrass biochar (Figure 3). Three of the seven maize biochars had K$_d$s less than 20 L kg^{-1} whereas the other four had values of 55 L kg^{-1} or greater. In general, the switchgrass biochars had lower

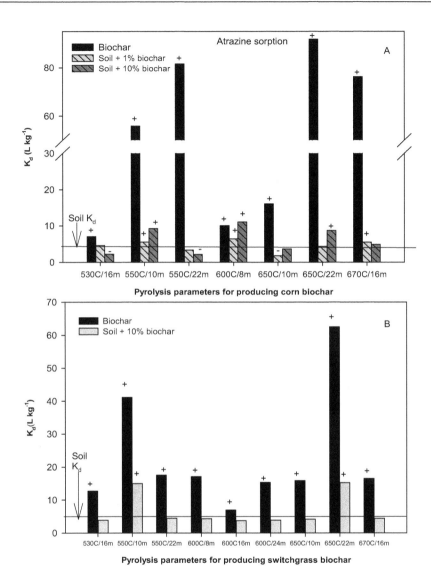

Fig. 3 A and B. Atrazine sorption (K_d) values to biochar from maize (*Zea mays*) stover (A) and switchgrass (*Panicum virgatum*) (B) produced by microwave pyrolysis at various processing times and temperatures. K_d values of sorption for the A horizon of a Brandt silty clay loam (Fine-silty, fmixed, superactive, frigid Calcic Hapludoll, [Soil Survey Staff, 2011]) soil when amended with 1 or 10% maize biochar or 1% switchgrass biochar. K_d sorption value of atrazine to unamended soil averaged about 3.86 L kg⁻¹. A "–" sign indicates lower sorption at $P \leq 0.05$ and a "+" sign indicates greater sorption at $P \leq 0.05$ than unamended soil.

K_d values for atrazine than maize, with only two of the nine samples having sorption values >18 L kg[-1]. Correlation analysis was conducted to examine pH of biochar vs K_d but these parameters were poorly to moderately correlated for maize (r = 0.4) and not correlated for switchgrass.

Amending soil with maize biochar at 1% increased the K_d with three biochars and decreased the K_d for one biochar. The maximum increase was 66% more sorbed than unamended soil. The 10% additions decreased the amount sorbed by soil in two samples by about 43%. This was surprising as one of the biochars alone had double the K_d of soil (K_d = 7 L kg[-1]) and a pH of 4.5 and the other had very high sorption (K_d = 82 L kg[-1]) value and pH of 7.6. It is unclear what properties of this biochar would result in lower atrazine sorption. The soil amended with three maize biochars used at 10% amendment had nearly 3 times as much atrazine sorbed (K_ds ranging from 8.7 to 11.0 L kg[-1]) when compared with soil alone. Two switchgrass biochars with the highest atrazine sorption also increased atrazine sorption when added as a 10% soil amendment, and raised the K_ds nearly 4-fold, with a K_d of about 15 L kg[-1]. Other switchgrass biochars had no or only a slight influence on atrazine sorption.

2.2.4 2,4-D sorption to biochar and soils amended with biochar

Unlike atrazine which has a positive charge in most soils, 2,4-D with a pKa of 2.8 is a weak acid in most soil solutions (Wauchope et al., 1992). This chemical was chosen as a model compound to explore the effect of biochar on these types of compounds. The negative charge on the 2,4-D, as well as other chemicals in this auxin-type chemistry, often results in low or no sorption to soil (Clay et al., 1988). If these types of chemicals have a long residence time in soil (e.g. picloram), there is a high potential for leaching, although, because 2,4-D often is reported to have a ½ life of 10 d or less, leaching of this chemical is not usually considered a problem.

The K_d sorption value of 2,4-D to unamended Brandt soil was about 1 L kg[-1,] a four-fold lower sorption than atrazine to this soil. All biochar samples had much greater sorption coefficients than soil alone (Figure 4), with switchgrass biochars generally sorbing more 2,4-D than maize biochars. The K_d values for all biochars, regardless of feedstock type ranged from about 3 to >80 L kg[-1] and was much greater than soil. K_d values for soil amended with 1% maize biochars were similar to K_d of unamended soil (Figure 4). Amending soil with 10% biochar (either maize or switchgrass) resulted in a few treatment combinations that had increased sorption compared to soil. Maize biochar resulting from processing stover at 600°C for 8 min increased 2,4-D sorption 3.3 times over unamended soils, whereas maize biochar formed from processing at 650°C for 22 min increased 2,4-D sorption by 4.5 times. Switchgrass biochar added at 10% to soil had little impact on 2,4-D sorption with two exceptions. The first was the biochar formed when processed at 550°C for 10 min where a 9.4- fold sorption increase was measured and the second when switchgrass was processed at 650°C for 22 min where a 15-fold sorption increase was measured. These two switchgrass biochars also dramatically increased atrazine sorption. The char produced at the higher temperature did influence soil EC values at 10% addition (Table 1), however, it is not known what the exact properties of these biochars or their interactions with soil/solution resulted in these increased sorption amounts.

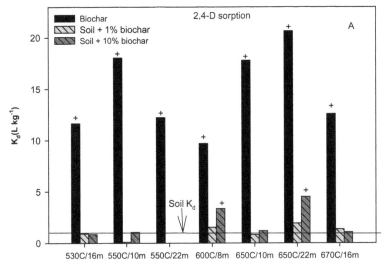

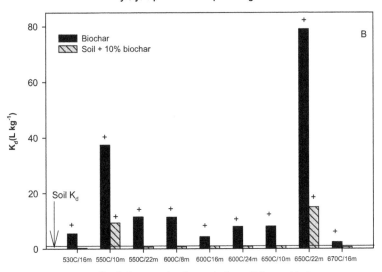

Fig. 4 A and B. 2,4-D sorption (K_d) values to biochar from maize (*Zea mays*)stover and switchgrass (*Panicum virgatum*) produced by microwave pyrolysis at various processing times and temperatures; K_d values of sorption for the A horizon of a Brandt silty clay loam (Fine-silty, mixed, superactive, frigid Calcic Hapludoll, [Soil Survey Staff, 2011]) soil when amended with 1 or 10% maize biochar or 10% switchgrass biochar. K_d sorption value of unamended soil averaged about 1.0 L kg⁻¹. A "+" sign indicates greater sorption at P≤ 0.05 than unamended soil.

3. Conclusion

Biochars, the by-products of pyrolitic conversion processes of vegetative biomass to gas, bio-oil, or other fuels, are proposed soil amendments for many diverse purposes. Biomass feedstocks and production processes vary depending on the desired end-products. This study measured the influence of several microwave pyrolitic conversion processes, which varied temperature and residence time, on pH and EC characteristics of the resulting biochars produced from maize stover and switchgrass. These biochars were used to amend a silty clay loam soil and examined the solution pH, EC, and sorption properties of a weakly cationic herbicide, atrazine, and an anionic herbicide, 2,4-D.

The microwave pyrolysis parameters of processing time and temperature of maize stover and switchgrass produced biochars that had a range of characteristics, with enough variation that they should not be thought of as a single entity with uniform properties. Short processing times (<10 min) of either feedstock at high (650°C) or low (550°C) temperature resulted in biochar with a pH < 4.5. Biochars produced with processing times >15 min at high temperature resulted in materials with pHs >8. Processing at intermediate temperatures and times resulted in char pHs ranging from 5.6 to 6.5. Adding 1% char to soil did not impact soil pH (6.4) whereas adding 10% biochar decreased soil pH a maximum of 12% when low pH biochars were used or increased soil pH up to 7% when high pH biochars were applied. "Native" soil EC was 1.63 mS/cm. Soils amended with 1% or 10% biochar ranged from -20% lower up to 39% higher EC values depending on biochar type and amount added. The biochars used in this study would be considered 'fresh', and not aged or post-process treated. Aging biochar or treating with steam or oxygen has been reported to dramatically change pH and other properties. Studies on these materials would need to be conducted to determine if results are similar to those reported for this study.

In a 2010 literature review, Kookana (2010) stated that there were limited published studies on the effect of biochars on pesticide efficacy and fate in soil, although in the few studies where sorption is reported, the sorption coefficients could be as high as >2000 times those of soil. Results from our study confirmed that when biochars were used as a single sorption material very high sorption amounts could be observed for both a cationic and an anionic compound. Herbicide sorption K_d to all biochars alone was very high compared with soil but varied among biochar types. Soil amended with 1% maize stover biochar had herbicide sorption values similar to unamended soil. However, adding 10% biochar amendment increased both atrazine and 2,4-D sorption coefficients by many-fold. A neutral herbicide, alachlor, has also been shown to have increased sorption in soils amended with woodchip biochar addition (Spokas et al., 2009). If biochars are applied to production fields, biochars may reduce atrazine preemergence weed control due to decreased availability to emerging seedlings. Kookana (2010) also discussed the possibility of longer residence time of pesticides due to reduced bioavailability, which may influence further the impact of a pesticide on ecotoxicology and potential accumulation. Indeed, Jones et al. (2011) reported biochar addition suppressed simazine biodegradation due to limiting availability to soil microbes through increased sorption, although leaching potential was reduced simultaneously.

The results of this study along with other reports have implications on best use of biochar in agricultural fields. If biochar has no or little effect on pesticide sorption, efficacy, or EC values, then the material may be suitable for general application in agricultural fields and highly desirable if it can be used to increase water holding capacity or as a nutrient source. Biochars, if high in sorption capacity, may be applied strategically and could accomplish important roles in ecosystem health and environmental quality. Biochar, added in filter strips and waterways, eroded landscapes, or other areas where increased sorption is desired, may aid in cleaning water running off fields by sorbing undesirable contaminants. Increased sorption may also slow or stop herbicides from leaching, so highly sorbent biochar types may be desired over shallow aquifers or in areas low in native organic matter (Wang et al., 2010). Herbicide bioavailability in some cases may be reduced, protecting sensitive plants.

Conversely, the effect of spreading biochars across entire fields may have negative results and be undesirable. One consequence may be that the materials increase soil EC values to saline levels. In addition, if the biochar reduces the efficacy of soil-applied herbicides or other pesticides this may have negative impacts. Reduced pesticide efficacy would require higher herbicide application rates to be as effective as lower rates. This would have monetary implications for growers and field managers by increasing management costs. Increased sorption, in some cases, also may increase the recalcitrance of pesticides leading to longer residence times in the environment. The occurrence of greater recalcitrance may be desirable if bioactivity was still acceptable and longer activity of the pesticide was desired to control the pest of interest. However, longer residence time may lead to other long-term environmental problems, such as greater leaching potential or carry-over problems into the following season.

Prior to any regular field applications of any biochar, the biochar properties must be examined to determine the suitability of the material for the long-term management of a particular site. The reasons for the application should be defined clearly and the outcomes closely monitored to determine if expectations and results are synonymous.

4. Acknowledgments

Funding provided by South Dakota Maize Utilization Council, US USDA/Sun Grant Initiative, and South Dakota Agricultural Experiment Station. Undergraduate participation included Mr. Mitch Olson, Mr. Dan Clay, and Ms. Kaitlynn Krack.

5. References

Abrol, I.P., Yadav, J.S.P., & Massoud, F. (1988). Salt affected soils and their management, Food and Agricultural Organization of the United Nations (FAO), Soils Bulletin 39.

Bridgewater, A.V. (2003). Renewable fuels and chemicals by thermal processing of biomass. *Chemical Engineering Journal*, 91, 2-3, (March 2003), pp. 87–102.

Bridgewater, A.V., Meier, D., & Radlein, D. (1999). An overview of fast pyrolysis of biomass. *Organic Geochemistry*, 30, 12, (December1999), pp. 1479–1493.

Brown, C.D., Carter, A.D., & Hollis, J.M. (1995). Soils and pesticide mobility, In: *Environmental Behaviour of Agrochemicals*, Roberts, T.R., & Kearney, P.C. (Eds.). pp. 131-184, John Wiley & Sons, Chichester, England.

Cao, X., Ma, L., Gao, B., & Harris, W. (2009). Dairy-manure derived biochar effectively sorbs lead and atrazine. *Environmental Science and Technology*, 43, 9, (May 2009), pp. 3285-3291.

Chidumayo, E. N. (1994). Effects of wood carbonization on soil and initial development of seedlings in miombo woodland, Zambia. *Forest Ecology and Management,* 70, 1-3, (December 1994), pp. 353–357.

Clay, S.A., Koskinen, W.C., Allmaras, R.R., & Dowdy, R.H. (1988). Differences in herbicide adsorption on soil using several soil pH modification techniques. *Journal of Environmental Science and Health B.* 23, 6, (1988), pp. 599-573.

Cox, D., Bezdicek, D., & Fauci, M. (2001). Effects of compost, coal ash, and straw amendments on restoring the quality of eroded Palouse soil. *Biology and Fertility of Soils,* 33, 5, (2001), pp. 365–372.

Daud, W.M.A.W., Ali, W.S.W., & Sulaiman, M.Z. (2001). Effect of carbonization temperature on the yield and porosity of char produced from palm shell. *Journal of Chemical Technology and Biotechnology,* 76, 12, (December 2001), pp. 1281–1285.

Demirbas, A. (2004). Effects of temperature and particle size on bio-char yield from pyrolysis of agricultural residues. *Journal of Analytical and Applied Pyrolysis,* 72, 2, (November 2004), pp. 243-248.

Erickson, C. (2003). Historical ecology and future explorations. In: *Amazonian Dark Earths: Origin, Properties, Management,* Lehmann, J., Kern, D.C., Glaser, B., &Woods, W.I. (Eds.), pp. 455–500, Kluwer Academic Publishers, Dordrecht, Netherlands.

Glaser, B., Haumaier, L., Guggenberger, G.& Zech, W. (2001). The Terra Preta phenomenon – a model for sustainable agriculture in the humid tropics. *Naturwissenschaften,* 88, 1, (January 2001), pp. 37-41.

Glaser, B., Lehmann, J., & Zech, W. (2002). Ameliorating physical and chemical properties of highly weathered soils in the tropics with charcoals-A review. *Biology and Fertility of Soils,* 35, 4, (June 2002), pp. 219–230.

Jha, P., Biswas, A.K., Lakaria, B.L., & Rao, A.S. (2010). Biochar in agriculture – prospects and related implications. *Current Science,* 99, 9, (November 2010), pp. 1218-1225.

Jones, D.L., Edward-Jones, G. & Murphy, D.V. (2011). Biochar mediated alterations in herbicide breakdown and leaching in soil. *Soil Biology & Biochemistry,* 43, 4, (April 2011), pp. 804-813.

Karaosmanoglu, F., Isigigur-Ergudenler, A., & Sever, A. (2000). Biochar from the straw-stalk of rapeseed plant. *Energy & Fuels,* 14, 2, (March-April 2000), pp. 336-339.

Katyal, S., Thambimuthu, K., & Valix, M. (2003). Carbonisation of bagasse in a fixed bed reactor: influence of process variables on char yield and characteristics. *Renewable Energy,* 28, 5, (April 2003), pp. 713–725.

Kishimoto, S., & Sugiura, G. (1985). Charcoal as a soil conditioner. In: *Symposium on Forest Products Research,* International Achievements for the Future, 5, (April 1985), pp.12–23 Republic of South Africa.

Kookana, R.S. (2010). The role of biochar in modifying the environmental fate, bioavailability, and efficacy of pesticides in soils: a review. *Australian Journal of Soil Research,* 48, special issue. 6-7, (June 2010), pp. 627-637.

Koskinen, W. C. & Clay, S.A. (1997). Factors Affecting Atrazine Fate in North Central U.S. Soils. *Reviews of Environmental Contamination and Toxicology,* 151, pp. 117-165.

Laird, D.A. (2008). The charcoal vision: A win-win-win scenario for simultaneously producing bioenergy, permanently sequestering carbon, while improving soil and water quality. *Agronomy Journal,* 100, 1, (January 2008), pp. 178-181.

Laird, D.A, & Koskinen, W.C. (2008). Triazine soil interactions. In: *The Triazine Herbicides. 50 years revolutionizing agriculture,* LeBaron, H.M., McFarland, J.E., & Burnside, O.C. (Eds.), pp. 275-299. Elsevier, Oxford, UK.

Lehmann, J. (2007). Bio-energy in the black. *Frontiers in Ecology and the Environment,* 5, 7, (September 2007), pp. 381-387.

Lehmann, J., Czimczik, C., Laird, D., & Sohi, S. (2009). Stability of biochar in the soil. In: *Biochar for Environmental Management,* Lehmann, J. & Joseph, S. (Eds.), pp.183-205, Earthscan, London, England.

Lehmann, C.J., da Silva Jr., J.P., Rondon, M., daSilva, C.M., Greenwood, J. Nehls, T., Stein, C., & Glaser, B. (2002). Slash-and-char - a feasible alternative for soil fertility management in the central Amazon? *Proceedings of 17th World Congress of Soil Science,* Bangkok, August, 2002.

Lehmann, C.J., & Rondon, M. (2006). Bio-char soil management on highly-weathered soils in the tropics. In: *Biological Approaches to Sustainable Soil Systems,* Uphoff, N.T. (Ed.), pp. 517-530, CRC Press, Boca Raton, FL.

Lei, H., Ren, S., & Julson, J. (2009). The effects of reaction temperature and time and particle size of maize stover on microwave pyrolysis. *Energy & Fuels,* 23, 6, (June 2009), pp. 3254-3261.

Mbagwu, J.S.C., & Piccolo, A. (1997). Effects of humic substances from oxidized coal on soil chemical properties and maize yield. In: *The Role of Humic Substances in the Ecosystems and in Environmental Protection,* Drozd, J., Gonet, S.S., Senesi, N., Weber, J. (Eds). pp. 921–925, IHSS, Polish Society of Humic Substances, Wroclaw, Poland.

McHenry, M. (2009). Agricultural bio-char production, renewable energy generation and farm carbon sequestration in Western Australia: certainty, uncertainty and risk. *Agriculture, Ecosystems and Environment,* 129, issues 1-3, (January 2009), pp. 1-7.

Oguntunde, P.G., Fosu, M., Ajayi, A.E., & van de Giesen, N. (2004). Effects of charcoal production on maize yield, chemical properties and texture of soil. *Biology and Fertility of Soils,* 39, 4, (March 2004), pp. 295-299.

Okimori, Y., Ogawa, M., & Takahashi, F. (2003). Potential of CO_2 emission reductions by carbonizing biomass waste from industrial tree plantation in Sumatra, Indonesia. *Mitigation and Adaptation. Strategies for Global Change,* 8, 3, (September 2003), pp. 261-280.

Piccolo, A., Pietramellara, G., & Mbagwu, J.S.C. (1996). Effects of coal derived humic substances on water retention and structural stability of Mediterranean soils. *Soil Use and Management,* 12, 4, (December 1996), pp. 209–213.

Richards, L.A. (ed.) 1954. Diagnosis and improvements of saline and alkali soils. USDA. Agriculture Handbook 60. 160 p.

Sanchez, M.E., Lindao, E., Margaleff, D., Martinez, O., & Moran, A. (2009). Pyrolysis of agricultural residues from rape and sunflower: production and characterization of biofuels and biochar soil management. *Journal of Analytical and Applied Pyrolysis,* 85, 1-2 (May 2009), pp. 142-144.

Sanchez, P.A., Villachia, J.H., & Bandy, D.E. (1983). Soil fertility dynamics after clearing tropical rainforest in Peru. *Soil Science Society of America Journal,* 47, 6, (November-December. 1983), pp. 1171–1178.

Skjemstad J.O., Reicosky, D.C., Wilts, A.R., & McGowan, J.A. (2002). Charcoal carbon in U.S. agricultural soils. *Soil Science Society of America Journal*, 66, 4, (July-August, 2002), pp. 1249-1255.

Sohi, S.P., Krull, E., Lopez-Capel, E., & Bol, R. (2010). A review of biochar and its use and function in soil. *Advances in Agronomy*. 105, (2010), pp. 47-82.

Sombroek, W., Ruivo, M.L., & Fearnside, P.M. (2003). Amazonian dark earths as carbon stores and sinks. In: *Amazonian Dark Earths: Origin, Properties, Management*, part 2 Lehmann, J., Kern, D. C., Glaser, B., & Woods, W. I. (Eds.), pp. 125-139, Kluwer Academic Publishers, Dordrecht, Netherlands.

Spokas, K.A., Koskinen, W.C., Baker, J.M., & Reicosky, D.C. (2009). Impacts of woodchip biochar additions on greenhouse gas production and sorption/degradation of two herbicides in a Minnesota soil. *Chemosphere*, 77, 4, (October 2009), pp. 574-581.

Soil Survey Staff, Natural Resources Conservation Service, United States Department of Agriculture. Official Soil Series Descriptions. Available online at http://soils.usda.gov/technical/classification/osd/index.html . Accessed August 24, 2011.

Tryon, E.H. (1948). Effect of charcoal on certain physical, chemical, and biological properties of forest soils. *Ecological Monographs*, 18, 1, (January 1948), pp. 81–115.

Uchimiya, M., Wartelle, L.H., Lima, I.M., & Klasson, K.T. (2010). Sorption of deisopropylatrazine on broiler litter biochars. *Journal of Agriculture and Food Chemistry*. 28, 23, (December 2010), pp. 12350-12356.

Wang, H.L., Lin, K.D., Hou, Z.N., Richardson, B., & Gan, J. (2010). Sorption of the herbicide terbuthylazine in two New Zealand forest soils amended with biosolids and biochars. *Journal of Soils and Sediments*, 10, 2, (March 2010), pp. 283-289.

Wauchope, R.D., Buttler, T.M., Hornsby, A.G., Augustijn-Beckers, P.W.M., & Burt, J.P. (1992) The SCS/ARS/CES pesticide properties database for environmental decision-making. *Reviews of Environmental Contamination and Toxicology*, 123, (1992) pp. 1–155.

Wauchope, R.D., Yeh, S., Linders, J.B.H.J., Kloskowski, R., Tanaka, K., Rubin, B., Katayama, A., Kördel, W., Gerstl, Z.,Lane, M., & Unsworth, J.B. (2002). Pesticide soil sorption parameters: theory, measurement, uses, limitations and reliability. *Pest Management Science*, 58, 5 (May, 2002), pp. 419-445.

Weber, J.B. (1995). Physicochemical and mobility studies with pesticides. In: *Agrochemical Environmental Fate: State of the Art*. Leng, M.L., Leovey, E.M.K., and Zubkoff, P.L. (Eds.), pp. 99-116, CRC Press. Boca Raton, FL.

Woolf, D., Amonette, J.E., Street-Perrott, F.A., Lehmann, J., & Joseph, S. (2010). Sustainable biochar to mitigate global climate change. *Nature Communications*, 1, 56, (August 2010), pp. 1-9.

Yaman, S. (2004). Pyrolysis of biomass to produce fuels and chemical feedstocks. *Energy Conversion and Management*, 45, 5, (March 2004), pp. 651–671.

Yu, X.-Y., Ying, G.-G., & Kookana, R. S. (2009). Reduced plant uptake of pesticides with biochar additions to soil. *Chemosphere*, 76, 5, (July 2009), pp. 665-671.

5

1(Heterocyclyl),2,4,5-Tetrasubstituted Benzenes as Protoporphyrinogen-IX Oxidase Inhibiting Herbicides

Hai-Bo Yu, Xue-Ming Cheng and Bin Li
State Key Laboratory of the Discovery and Development of Novel Pesticide
Shenyang Research Institute of Chemical Industry Co. Ltd.
China

1. Introduction

It is well known that agrochemicals have played a important role in agricultural production that provided for about 700 million people during the past 50 years. At the same time, the increasing world population seems to be a major driving force for the need to enhance the output of food production. The agrochemical industry has been very successful in developing new herbicides and other agrochemicals. Herbicides are used widely in the world in protecting crops from undue competition from weeds (Price & Kelton, 2011).

The first commercial inhibitor of protoporphyrinogen oxidase (Protox) is the nitrofen that belongs to diphenyl ether (DPE), which was introduced in 1963 by Rohm & Hass (Now Dow AgroSciences) (Matsunaka, 1976). Some years later, the oxadiazon as the fisrt compound of the 1(heterocyclyl), 2, 4, 5-tetrasubstituted benzene (HTSB) family was introduced in 1968 by Rhone-Poulenc (Metivier et al., 1968). Nitrofen and oxadiazon represent the earliest examples of Protox inhibiting herbicides (Fig. 1). Although their chemical structures are completely different from each other, they share a common mode of action, inhibition of the protoporphyrinogen oxidase enzyme, though this was not known until the late 1980s.

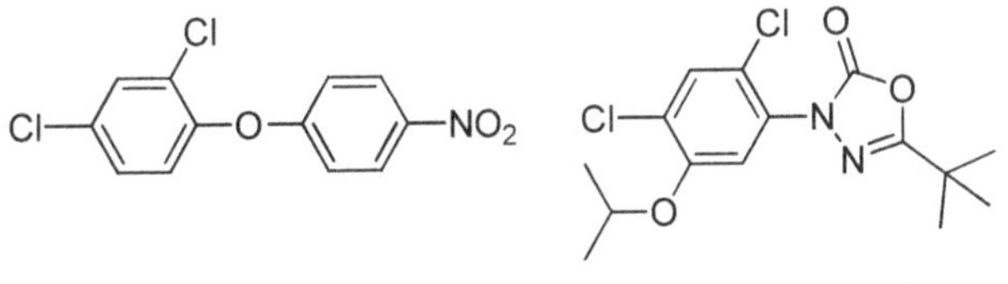

nitrofen 1963   oxadiazon 1968

Fig. 1. Chemical structures of two early examples of Protox inhibitors.

Several early inventions of HTSB herbicide in 1960s had a significant impact on our understanding of the structure–activity of this kind herbicides. Rhone-Poulenc first introduced 3-(2,4-dichlorophenyl)-1,3,4-oxadiazol-2(3H)-one in 1965 (Boesch et al., 1965). Further lead optimization at the phenyl ring soon led to the discovery in 1968 of the 2,4-dichloro-5-isopropoxyphenyl substitution pattern of the herbicide oxadiazon

(Boesch et al.,1968). Oxadiazon was the first compound of the cyclic imide family introduced into the markert for the control of annual grasses and broadleaf weeds in pre-emergence or early post-emergence treatment by Rhone-Poulenc in 1969. The second cyclic imide herbicide, chlorophthalim, was introduced by Matsui in 1972. The 2, 4-dihalo-5-substituted pattern at the aromatic ring would become the basis for much of the 2,4,5-trisubstituted phenyl tetrahydrophthalimide research that followed in this area of chemistry.

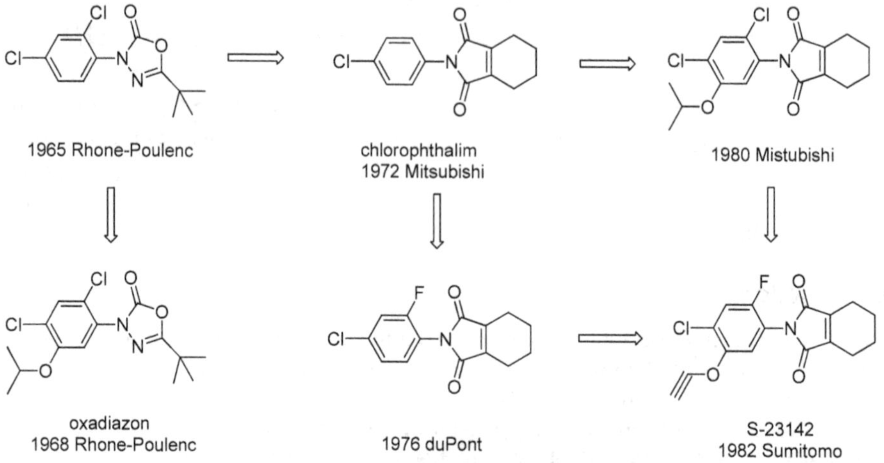

1965 Rhone-Poulenc   chlorophthalim   1980 Mistubishi
                            1972 Mitsubishi

oxadiazon   1976 duPont   S-23142
1968 Rhone-Poulenc   1982 Sumitomo

Fig. 2. Incorporation of the 2,4-dihalo-5-alkoxy aromatic pattern of oxadiazon into new phenyl tetrahydrophthalimide ring systems.

A breakthrough discovery was the increasing biological activity caused by the replacement of chlorine by fluorine at the 2-phenyl position. In 1976, DuPont introduced the first example of a 2-fluoro-4-chlorophenyl tetrahydrophthalimide Protox inhibitor (Goddard, 1976) (Fig. 2). The dramatic increase in biological activity caused by the fluorine in the 2 position of the phenyl ring would, in the next decade, the 1980s, influence the lead optimization work in the HTSB area, such as the discovery of the 4-chloro-2-fluorophenyltetrahydrophthalimide herbicide S-23142 (Nagano et al., 1982). Since then, various HTSB derivatives with high herbicidal activity (10-50 g/ha) have been discovered by many companies.

2. Commercialized protoporphyrinogen-IX oxidase inhibiting herbicides

A number of Protox inhibiting compounds have been already commerciallized. Their structures and biological activities are introduced as follows.

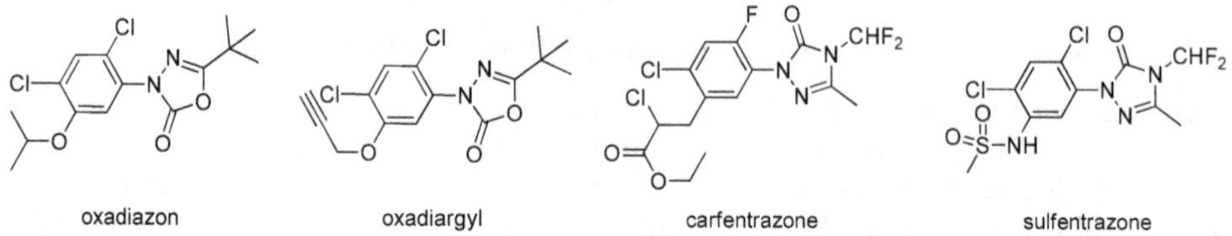

oxadiazon   oxadiargyl   carfentrazone   sulfentrazone

Fig. 3. Chemical structure of oxadiazon, oxadiargyl, carfentrazone, sulfentrazone.

Both oxadiazon and oxdiargyl are developed by Rhone-Poulenc (Fig. 3). Oxadiargyl is a pre-emergence and early post-emergence herbicide in sugarcane and sunflower fields or orchards to control both annual broadleaf and grass weeds. It application rate is rather high at 500-2000 g a. i. /ha. It is also developed as an herbicide for rice and turf.(Oe et al., 1995). FMC developed carfentrazone-ethyl and sulfentrazone. Carfentrazone-ethyl is a post-emergence herbicide in rice, cereals and maize fields. It shows excellent activity against annual broadleaf weeds such as *Gallium, Lamium,* and *Veronica* in wheat at 20-35 g a. i./ha. It also controls *Euphorbia, Polygogum, Abutilon, Ipomea, Kochia, Salsola,* etc. in foliar application (Vansaun et al., 1993; Mize, 1995). Sulfentrazone is a pre-emergence herbicide in soybean and sugarcane fields. It controls *Ipomea, Amaranthus, Chenopodium, Abutilon, Polygonum, Datura,* etc. at 350-420 g a. i. /ha. (Oliver et al., 1996; Vidrine et al., 1996)

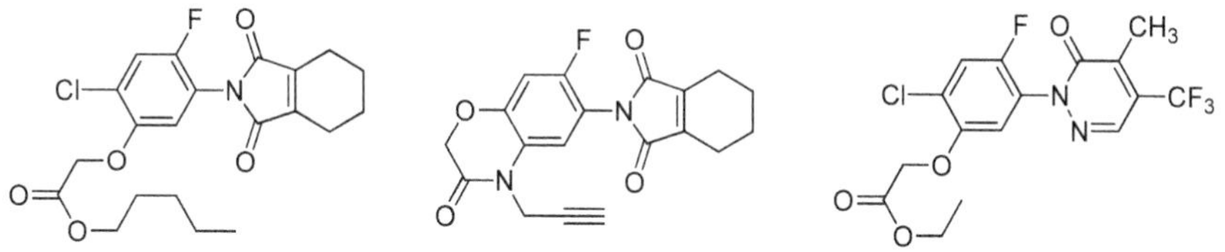

flumiclorac pentyl   flumioxazin   flufenpyr-ethyl

Fig. 4. Chemical structure of flumiclorac pentyl, flumionazin, flufenpr-ethyl.

The above three herbicides in figure 4 are all developed by Sumitomo. Flumiclorac pentyl was commercialized as the first cyclic imide herbicide in Europe as early as 1993. It controls annual broadleaf weeds such as *Abutilon, Euphorbia, Chenopodium, Datura, Ambrosia* and *Xanthium* at 30-60 g a. i. /ha in post- emergence treatment in soybean and maize fields, especilly, it shows excellent activity against *Abutilon at progressed leaf stage at 60* g a. i. /ha (Kurtz & Pawlak, 1992, 1993; Satio et al., 1993). Flumioxazin controls annual broadleaf weeds such as *Abutilon, Euphorbia, Chenopodium, Ipomea* and *Sida* at 50-100 g a. i. /ha in pre-emergence treatment in soybean and peanut fields. But it is less active against annual grass weeds at the same dosage rate.(Yoshida et al., 1991). Flufenpyr-ethyl is commercillized recently as an herbicide in soybean, cotton, corn and sugarcane. It is supposed that it is more selective.

Pentoxazone was discovered by Sagami Chemical Research Center and Kaken Pharmaceutical as a pre-emergence and early post-emergence herbicide in rice (Fig. 5.). It shows excellent activity against both annual and perennial weeds such as *Echinochloa, Eleochairs, Sagittaria* and *Cyperus* at 145-150 g a. i. /ha (Yoshimura et al., 1992; Hirai et al., 1995). Pyraflufen-ethyl was developed by Nichino, It shows excellent selectivity for winter cereals at the extremely low rates of 6 to 12 g /ha, and also long-term residual activity brought out by chemical and biological stabilities , as a post-emergent contact herbicide active on broadleaf weeds such as cleavers, henbit, clickweeds and wild chamomile; especially provided effective control of 5- to 6-leaf stage of cleavers at the low rate. Azafenidin is developed by Du Pont as a non-selective pre- and post-emergence herbicide for non-cropland and orchard. It has wide herbicidal spectra and controls both annual and perennial weeds at 560 g a. i. /ha (Netzer et al., 1996). Saflufenacil was developed by BASF as an herbicide which is used alone or in mixtures with glyphosate for burn down weed control, with foliar and residual activity against more than 70 broadleaf weeds, introduced in Nicaragua, Chile, and Argentina as Heat in 2009.

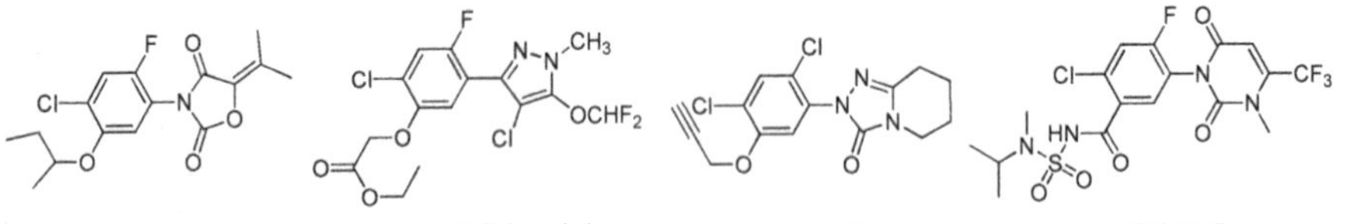

Fig. 5. Chemical structure of pentoxazone, pyraflufen-ethyl, azafenidin, saflufenacil.

3. Mode of action of 1(heterocyclyl),2,4,5-tetrasubstituted benzenes protoporphydrinogen-IX oxidase inhibiting herbicides

The mode of action of 1(heterocyclyl),2,4,5-tetrasubstituted benzenes (HTSB) protoporphyrinogen oxidase (Protox) inhibiting herbicides has been extensively reviewed (Duke et al., 1990). HTSB herbicides inhibit the enzyme Protox in the chlorophyll biosynthesis pathwy (Matringe & Scalla, 1988; Witkowski & Halling, 1988; Lydon & Duke, 1988). The Protox enzyme catalyzes the oxidation of protoporphyrinogen IX to protoporphyrin IX by molecular oxygen. Inhibiting the Protox enzyme, which is located in the chloroplate envelop, results in an accumulation of the enzyme product protoporphyrinogen IX, but not the substrate, via a complex process that has not been entirely elucidated. Enzymatic oxidation of protoporphyrin IX in the cytoplasm yeilds a significant accumulation of protoporphyrin IX from the location of the chlorophll biosynthesis, sequence in choroplasts. In the presence of light, protoporphyrin IX generates large amounts of singlet oxygen ($^1O_2$), which results in the peroxidation of the unsaturated bonds of fatty acids found in cell membranes (Fig. 6). The end result of this peroxidation process is the loss of membrane integrity and leakage, pigment breakdown, and necrosis of the leaf that results in the death of the plant. This is a relatively fast process, with leaf symptoms such as a flaccid wet appearance observed within hours of plant exposure to the Protox herbicides under sunlight.

4. Structure-activity relationships of 1(heterocyclyl),2,4,5-tetrasubstituted benzenes protoporphydrinogen-IX oxidase inhibiting herbicides

4.1 Overviw of structure-activity relationships of 1(heterocyclyl),2,4,5-tetrasubstituted benzenes protoporphydrinogen-IX oxidase inhibiting herbicides

It is very important to analyze structure-activity data accumulated during past trials when formulating rationnal structure-activity relationships (SARs). The relationships could be utilized as possible guiding principles for further structure transformation leading to novel peroxided herbicides. The information about (sub)molecular mechanisms of biological action may be extracted from the relationship. The structure-activity of Protox herbicides has been extensively reviewed (Fujita & Nakayama, 1999). Figure 6 shows the SARs of 2-fluoro-4-chloro-5-substituted-phenyl heterocycles (Theodoridis, 1997). SAR studies of 2,4,5-trisubstituted-phenyl heterocycles have shown that position 2 of the phenyl ring required a halogen group for optimum biological activity, with fluorine generating the highest overall activity. Introducing a substituent in position 3 of phenyl ring resulted in dramatic decrease

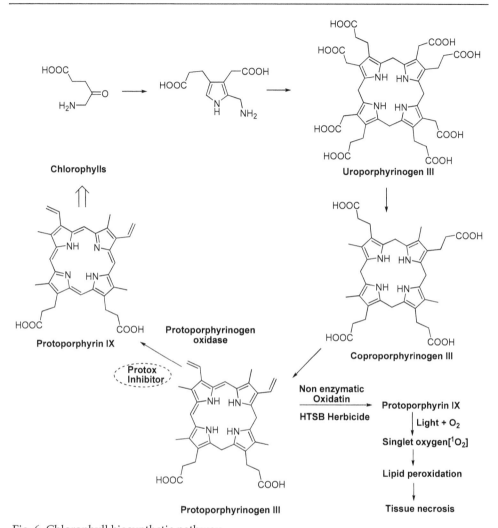

Fig. 6. Chlorophyll biosynthetic pathway.

of herbicidal activity. Position 4 of the phenyl ring required a hydrophobic, electronegative group such as halogen for optimum activity, with chlorine resulting in the best activity. Electron-donating groups such as methoxy resulted in significant loss of biological activity. As shown in Figure 7, the substituent R have a great effect on the bioactivity. Considering herbicidal activity and limited crop selectivity, OCH_2CCH is more favorable than other sunstituents. Considering weed spectrum and multicrop selectivity, R is $NHSO_2Et$ generating the highest overall activity. Many heterocyclic systems, usually attached to aromatic rings via a nitrogen atom, have been introduced in the past fifteen years. Oxadiazolinone (Metivier et al., 1968), oxazolidinedione (Hirai et al., 1989), tetrahydrophthalimide (Matsui et al., 1972), tetrazolinone (Theodoridis et al., 1990), triazolinone (Theodoridis, 1989), pyrimidinedione ring (Wenger, et al., 1988) showed relative higher herbicidal activity than other heterocyclic systems.

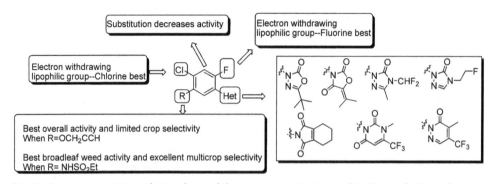

Fig. 7. Structure-activity relationships of the two aromatic rings of 2,4,5-trisubstituted-phenyl heterocyclic systems.

Linking the 4 and 5 positions of phenyl ring to give a new benzoheterocyclic ring, such as benzoxazinone, quinolin-2-one, benzimidazole, resulted in two new classed of Protox herbicides, both increased biological efficacy and new SARs (Fig. 8) (Lyga et al., 1999; , Grawford et al., 1997). As privious studies, position 2 of the phenyl ring required a halogen group for optimum biological activity, with fluorine generating the highest overall activity. The substituent R has a great effect on the herbicidal activity. Introducing propargyl resulted in dramatic increase in the bioactivity.

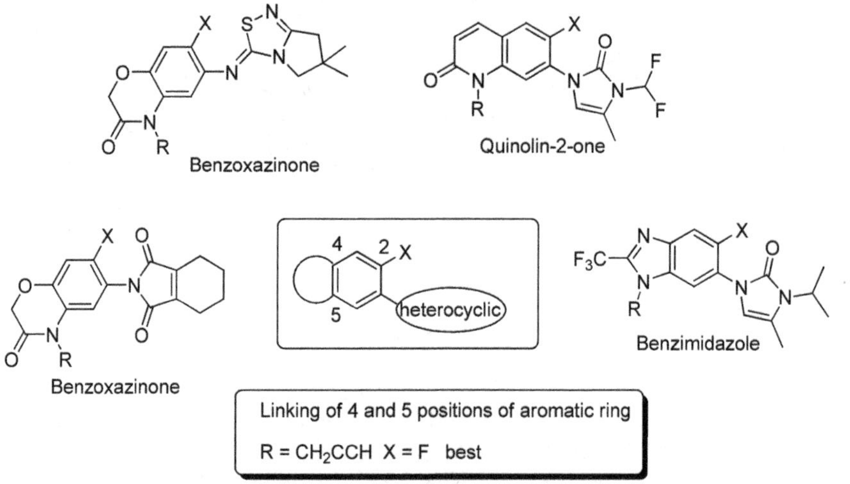

Benzoxazinone

Quinolin-2-one

Benzoxazinone

4 2 X

5 heterocyclic

Benzimidazole

Linking of 4 and 5 positions of aromatic ring

R = CH$_2$CCH X = F best

Fig. 8. Structure-activity relationships of benzoheterocycles resulting from linking aromatic positions 4 and 5 of phenyl heterocyclic systerms.

The second class of benzoheteroaryl Protox herbicides are obtained when aromatic positon 5 and 6 are linked togther to form various benzoherocyclic rings, which can be attached to a wide range of heterocycles (Fig. 9). The 6-trifluoromethyl group (R_1=CF$_3$) in the uracil ring is essential for bioactivity, replacing it with methyl results in complete loss of activity (Thecodoridis et al., 2000 and 1994). Increaseing the size and length of R_2 group resulted in a significant reduction in bioactivity. Substituents X had a dramatic effect on the weed spectrum and crop selectivity. Compounds with fluoroine and hydrogen resulted in broad-spectrum control of weeds and high herbicidal activity (Lyga et al., 1999).

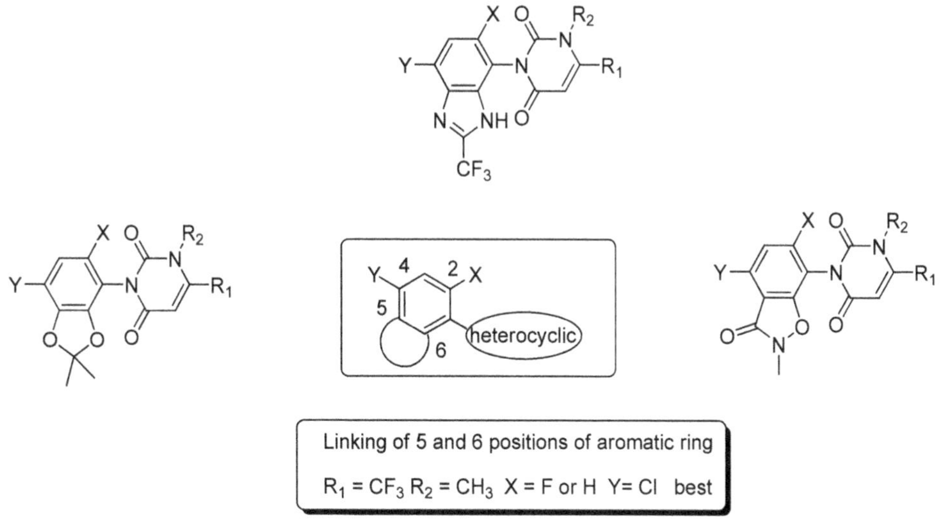

Fig. 9. Structure-activity relationships of benzoheterocycles resulting from linking aromatic positions 5 and 6 of phenyl heterocyclic systerms.

4.2 Pharmacophore analysis

Many molecular modeling studies of ligand binding require some knowledge of the pharmacophore as a starting point. The pharmacophore mode could show the significant structural similarities and identifies the active conformation. The model itself identifies the list of feature classes required and the distances between them. Eight HTSB compounds in table 1 was selected as the testing compounds for the pharmacophore study. A DISCO model of the pharmacophore was developed based on information from X-ray crystal structures of compound I-1 and Sybyl using the Tripos force field. Key pharmacophore elements are a polarizable functionality separated by a fixed distance from two H-bond accepting elements. The compound I-1 was choosed as a reference compound. The crystal data was listed in table 2 and the structure was shown in figure 10. The 3D structures of all the compounds in table 1 were built by SYBYL 6.9/ Sketch, and then optimized using MMFF94 force field, by powell method with energy termination of 0.005 kcal/mol, and a maximum of 1000 iterations. Then, the Gasteiger–Hückel charges were added. Compounds I-1 was selected as the training set and the test set.

As shown in figure 11, the pharmacophore model contains two hydrophobic centers and two acceptor atoms. One hydrophobic center was closed to the 1-heterocyclic ring, which may be interacted with phenylalanine 392 in Protox. Another hydrophobic center, which was in the center of phenyl group, may be interacted with leucine 356 and 372 in Protox. The acceptor atom was oxygen atom located in the carbonyl group and the 5-positon of phenyl group, respectively. The acceptor atom may have interaction with hydrogen atom in target enzyme.

Compound	Structure	PPO pI_{50}	Compound	Structure	PPO pI_{50}
I-1		8.97	I-5		8.49
I-2		8.86	I-6		8.57
I-3		8.92	I-7		8.49
I-4		8.55	I-8		8.96

Table 1. The HTSB compounds for pharmacophore study.

Formula	C17 H13 Cl F N O3
Formula weight	333.73
Color/shape	colorless/prism
Crystal system	Triclinic
Space group	P-1
Unit cell dimensions	a = 9.313(5) Å alpha = 99.663(9) deg.
	b = 9.380(5) Å beta = 104.263(9) deg.
	c = 10.485(6) Å gamma = 112.778(8)
Volume	deg.
Z	782.1(8) Å3
Calculated density	2
Absorption coefficient	1.417 g.cm-3
F(000)	0.269 mm-1
Crystal size/mm	344
Temp. /K	0.36 x 0.24 x 0.22
θ ranges/°	293(2)
Limiting indices	2.46 to 25.00 deg.
Reflections collected / unique	-11<=h<=9, -9<=k<=11, -10<=l<=12
Completeness to theta = 25.00	4071 / 2747 [R(int) = 0.0174]
Absorption correction	99.4 %
Max. and min. transmission	Semi-empirical
Refinement method	1.000000 and 0.689808
Data / restraints / parameters	Full-matrix least-squares on F2
Goodness-of-fit on F2	2747 / 1 / 208
Final R indices [I>2sigma(I)]	1.022
R indices (all data)	R1 = 0.0428, wR2 = 0.1095
Largest diff. peak and hole	R1 = 0.0632, wR2 = 0.1228
	0.440 and -0.291 e.A-3

Table 2. Crystal data and structure refinement of compound I-1.

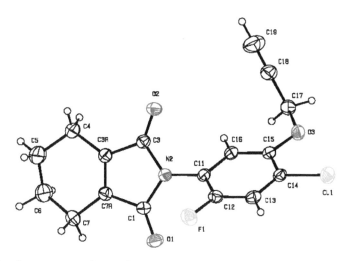

Fig. 10. Molecular structure of crystal compound I-1.

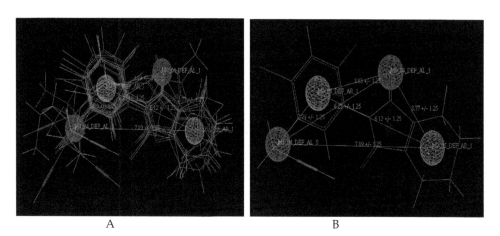

| A | B |

Fig. 11. Pharmacophore A: overlap of all eight compounds; B: model with referecce compound I-1.

4.3 CoMFA analysis

QSAR (quantitative structure-activity relationships) have been reported about the effects on the substituents in phenyl group (Fujita & Nakayama, 1999; Jiang et al., 2010), but little literature about comparative molecular field analysis (CoMFA) on the effect of heterocyclic rings (Zhang et al., 2011). In order to understand the heterocyclic rings and substituents in phenyl group effects on the PPO inhibition of a series of cyclic imide compounds, the method of CoMFA was applied to understand the quantitative structure–activity relationships. Using the information help us to increase the efficiency of syntheses of new candidate compounds.

37 HTSB compounds in table 3 was selected for the CoMFA study (Fujita & Nakayama, 1999). The 3D structures of all the compounds were built by SYBYL 6.9/ Sketch, and then optimized using MMFF94 force field, by Powell method with energy termination of 0.005 kcal/mol, and a maximum of 1000 iterations. Pharmacophore-based molecule alignment method was applied to superimpose all the compounds by using GALAHAD in SYBYL 6.9. The steric and electrostatic field energies for CoMFA were calculated using the SYBYL default parameters: 2.0 grid points spacing, a sp^3 carbon probe atom with +1 charge and a van der Waals radius of 1.52, and column filtering of 2.0 kcal/mol. The CoMFA descriptors were used as independent variables, and pI$_{50}$ values were used as dependent variables in partial leastsquares (PLS) regression analyses to derive 3D-QSAR models. Leave-one-out (LOO) cross-validated PLS analyses were performed to determine the optimal number of components to be used in the final QSAR models and to check the predictive ability of the models. To visualize the 3D-QSAR results in term of field contributions, isocontour maps were generated using the field type 'stdev * coeff' and the contour levels were set to default values. In CoMFA, compounds II-1 was selected as the training set and the test set.

The alignment of HTSB compounds was shown in figure 12. The molecular modeling studies found good overlap between the 37 HTSB compounds. As listed in Table 3, a predictive CoMFA model was established with the conventional correlation coefficient R^2 = 0.908, the standard error s = 0.319, and F-test value F = 49.5. The contribution of steric and electrostatic fields are 49.5% and 50.5%, respectively. The observed and calculated activity values for all the compounds are given in Table 3, and the plots of the caculated versus the actual activity values for all the compounds are shown in Figure 13.

In Figure 14, the isocontour diagrams of the steric and electrostatic field contributions ("stdev*coeff") obtained from the CoMFA analysis are illustrated together with exemplary ligands. The steric field contour map is plotted in Figure 14A. The green region highlights positions where a bulky group would be favorable for higher PPO inhibition activity. In contrast, yellow indicates positions where a decrease in the bulk of the desired compounds is favored. As shown in Figure 14A, the CoMFA steric contour plots indicated that a big yellow region is located around the group of phenyl in 4, 5-position, while a big green region surrounded the heterocycle group. This map means that the substituents of phenyl in 4, 5-position should be bulky. This steric map explained clearly why compound II-20 and II-21 displayed lower activity than other compounds. The electrostatic contour plot is shown in Figure 14B. The blue contour defines a region where an increase in the positive charge will result in an increase in the activity, whereas the red contour defines a region of space where increasing electron density is favorable. As shown in Figure 14B, the electrostatic contour plot showed that a blue region is around the Z group in the position of 5 of phenyl ring, whereas a red region is around the carbonyl group. The electrostatic contour plot indicated the target compounds bearing an electron-withdrawing group at the position of 5 of phenyl will display higher activity. This contour map indicated that the more electronegative of the oxygen atom of the carbonyl, the higher the activity of inhibitors. This means the carbon atom of one of the carbonyl group played an important role in

Compound	Structure	X, Y, Z	PPO-pI$_{50}$ Obs.	PPO-pI$_{50}$ Cal.	Deviation
II-1		2-F, 4-Cl, 5-OCH(Me)CCH	9	8.69	0.31
II-2		2-F, 4-Cl, 5-OCH$_2$CCH	8.97	9.04	-0.07
II-3		4-Cl, 5-COO-i-Pr	8.9	8.69	0.21
II-4		2-F, 4-Cl, 5-COO-i-Pr	8.86	8.82	0.04
II-5		2-F, 4-Br	8.6	8.51	0.09
II-6		2-F, 4-Cl, 5-OMe	8.52	8.76	-0.24
II-7		2-F, 4-Cl	8.43	8.54	-0.11
II-8		4-Cl, 5-COOEt	8.43	8.17	0.13
II-9		4-Cl, 5-COOCH$_2$COOMe	8.3	8.36	-0.06
II-10		4-Cl, 5-COOMe	8.05	8.12	-0.07
II-11		2-F, 4-Cl, 5-OCHF$_2$	7.96	7.95	0.01
II-12		4-Cl, 5-COO-t-Bu	7.85	8.03	-0.18
II-13		4-Br	7.67	6.80	0.87
II-14		4-Cl	7.6	6.83	0.77
II-15		4-OMe	7	7.11	-0.11
II-16		4-CF$_3$	6.55	6.57	-0.02
II-17		4-NO$_2$	6.4	6.62	-0.22
II-18		4-F	6.37	6.97	-0.60
II-19		4-Me	6.08	6.05	0.03
II-20		H	5.8	6.55	-0.75
II-21		3-Cl	5.6	5.76	-0.16
II-22		W=S; 2-F, 4-Cl, 5-OCH2C*CH	9.05	9.09	-0.04
II-23		W=S; 2-F, 4-Cl, 5-O-i-Pr	8.92	8.88	0.04
II-24		W=S; 4-Cl	8.17	8.32	-0.15
II-25		W=S; 2-F, 4-Cl	8.14	8.12	0.02
II-25		W=S; 4-Br	8.1	8.12	-0.02
II-27		W=O; 2-F, 4-Cl, SCH$_2$COOMe	8.08	8.08	0.0041
II-28		2-F, 4-Cl, 5-OCH(Me)C*CH	8.57	9.02	-0.45
II-29		4-Cl	7.66	7.32	0.34
II-30		2-F, 4-Cl, 5-OCH$_2$C*CH	9.14	9.24	-0.10
II-31		2-F, 4-Cl, 5-OCHF$_2$	8.55	8.46	0.09
II-32		OCH$_2$COOEt	8.49	8.40	0.09
II-33		H	8.49	8.51	-0.02
II-34			8.29	8.17	0.12
II-35			7.77	7.81	-0.04
II-36			8.49	8.42	0.07
II-37			8.96	8.79	0.17

Table 3. CoMFA study on the HTSB compounds.

determining the activity of PPO inhibitors. The stronger the ability of the carbonyl group to accept electrons from receptor, the higher the activity of PPO inhibitor.

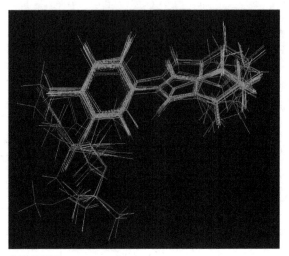

Fig. 12. Alignment of 37 HTSB compounds.

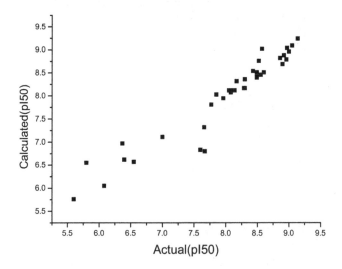

Fig. 13. Calculated pI_{50} (Y-axis) are versus actual pI_{50} (X-axis) values. The dots represent training compounds.

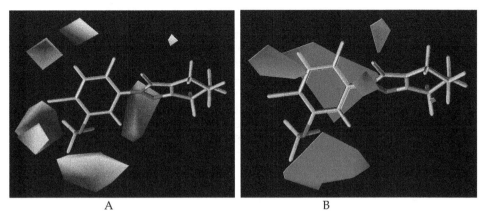

A   B

Fig. 14. CoMFA contour maps with compound II-1 as the reference structure. (A) Steric contours. Scattered green areas are regions where bulky substituents are favorable, yellow areas are unfavorable. (B) Electrostatic contours. The red areas are the regions where negative potential is favorable for the activity, blue areas are unfavorable.

5. Conclusions

HTSB herbicides are characterised with their high herbicidal activities, fast acting, and environmentally benign. However, most of them cause short-term damage to the crops applied, which makes them no significant market share in the last 30 years (Qasem, 2011).

In recent years, the development of Protox inhibitor-resistant crops (Li and Nicholl, 2005; Vencill, 2011) began a new era for the use of Protox herbicides. Furthermore, weed shifts observed in genetically modified crops, caused by the development of weed resistance to the widely used glyphosate herbicide, will offer market opportunities for herbicides with other modes of action, such as Protox inhibiting herbicides. So, Protox inhibiting herbicides will continue to be an important area of interest to agrochemical companies, with most efforts focused on fine tuning the 5 position of the phenyl ring. The application of CoMFA approach will seed up the discovery processes.

6. Acknowlegments

This project was supported by the National Basic Research Program of China (973 Program) (No.2010CB735601) and National Key Technology Support Program during the 12th Five-Year Plan Period (No. 2011BAE06B03-03). We would like to thank Professor Huazheng Yang (Nankai University) for the guidance and help in the Pharmacophore and CoMFA analysis.

7. References

Amália Jurado, A.S,; Fernandes, M.; Videira, R.; Peixoto, F. & Vicente, J. (2011). Herbicides: the Face and the Reverse of the Coin. An in Vitro Approach to the Toxicity of Herbicides in Non-Target Organisms, In: *Herbicides and Environment*, Kortekamp A. (Ed.), InTech, pp. 5-12, Available from:

http://www.intechopen.com/articles/show/title/herbicides-the-face-and-the-reverse-of-the-coin-an-in-vitro-approach-to-the-toxicity-of-herbicides-i

Boesch, R. & Metivier, J. (1965). New derivative of ioxadiazolone preparation. FR patent, FR 1394774.

Boesch, R. & Metivier, J. (1968). Oxadiazoline compounds and herbicidal compositions containing them. GB patent, GB 1110500.

Duke, S.O.; Becerril J.M.; Sherman, T.D.; Lydon, J. & Matsumoto, H. (1990). The role of protoporphyrin IX in the mechanism of action of diphenyl ether herbicides. *Pestic. Sci.* Vol. 30, pp. 367-378, ISSN: 1096-9063

Fujita, T. & Nakayama, A. (1999).Structure-activity relationship and molecula design of peroxidizing herbicides with cyclic imide structures and their relatives, in Peroxidizing herbicides, ed. Boger P and Wakabayashi K., pp. 92-115, Springer-Verlag, ISBN: 3-540-64550-0, Berlin

Goddard, S.J. (1976). Herbicidal isoindol-1-one derivatives. US Patent, US 3,992,189

Grawford, S.D.; Maravetz, L.L. & Thecodoridis, G. (1997). Herbicidal 3-(bicyclic nitrogen-containing heterocycle)-substituted-1-methyl-6-trifluoromethyluracils US patent, US 5,661,108

Hirai, K.; Futikami, T.; Murata, A; Hirose, H. & Yokata, M. (1989). Oxazolidinedione derivatives, method of producing the derivatives, and herbicides containing the derivatives. US patent, US 4,818,272

Hirai, K.; Yano, T.; Ugai, S.; Yamada, O.; Yoshimura, T.& Hori, M. (1995). A new low-rate pre-emergence herbicide KPP-314 for rice. *Proc. 15th Conf. Asian Pacific Weed Sci. Soc.*, Vol. 39, pp. 840-845

Jiang, L.L.; Tan, Y.; Zhu, X.L.; Wang, Z.F.; Zuo, Y.; Chen, Q.; Xi, Z. & Yang, G.F. (2010). Design, synthesis and 3D-QSAR analysis of novel 1,3,4-oxdiazol-2(3H)-ones as protoporphyrinogen oxidase inhibitors. *J. Agric. Food Chem.*, Vol. 58, pp. 2643-2651, ISSN: 1520-5118

Kurtz, A.R.& Pawlak, J.A. (1992). Postemergence weed control in field corn with V-23031 Herbicide *Proc. North Central Weed Sci. Soc.*, Vol. 47, pp. 47

Kurtz, A.R. & Pawlak, J.A. (1993). V-23031 – a new post emergence herbicide for use in field corn. *Abstr. Weed Sci. Soc. Am.*, Vol. 33, pp. 9

Lydon, J. & Duke, S.O. (1988). Porphyrin synthesis is required for photobleaching activity of the p-nitrosubstituted diphenyl ether herbicides. *Pestic. Biochem. Physiol.,*Vol. 31, pp. 74-83, ISSN: 0048-3575

Lyga, J.W.; Chang, J.H.; Theodoridis, G. & Baum, J. S. (1999). Structural replacements for the benzoxazinone protox inhibitors. *Pestic. Sci.* Vol. 55, pp. 281-287, ISSN: 1096-9063

Li, X.G. & Nicholl, D. (2005). Development of PPO inhibitor-resistant cultures and crops. *Pestic. Manag. Sci.*, Vol. 61, pp 277-285, ISSN: 1096-9063

Matringe, M. & Scalla, R. (1988). Effects of acifluorfen-methyl on cucumber cotyledons: Porphyrin accumulation. *Pest. Biochem. Physiol.*, Vol. 32, pp. 164-172, ISSN: 0048-3575

Matsui, K.; Kasugai, H.; Matsuya, K. & Aizawa, H. (1972). N-substituted-tetrahydro-phthalimides. FR patent, FR 2,119,703

Matsunaka, S. (1976). Diphenyl ethers . In : Kearney, P.C.; Kaufman, D.D. Herbicides: Chemistry degradation, and mode of action, vol2, Dekker, New York, pp. 709-739. ISBN-10: 0824780574

Metivier, J. & Boesch, R. (1968). 5-t-butyl-3-phenyl-2-oxadiazolones. US Patent, US 3,385,862

Mize, T.W. (1995). Control of annual weeds with F8426 in small grains. *Proc. Southern Weed Sci. Soc.*, Vol. 48, pp. 83-84

Nagano, E.; Hashimoto, S.; Yoshida, R.; Matsumoto, H.; Oshio, H. & Kamoshita, K. (1982). Tetrahydrophthalimides, and their production and use. EP patent, 61741

Netzer, D.A.; Riemenschneider, D.E. & Bauer, E.O.(1996). Pre and post flush application of Du Pont R6647 in hybrid polar plantations. *Proc. North Central Weed Sci. Soc.*, Vol. 51, pp. 105

Oe,Y.; Kawaguchi, S.; Yokoyama, M. & Jikihara, K. (1995). RP020630- A new herbicide for transplanted for rice(Oryza sative). *Proc. 15th Conf. Asian Pacific Weed Sci. Soc.* pp. 239-243

Oliver, L.R. Swantek, J.M. & King, C.A. (1996). Weed control in soybeans with sulfentrazone. *Abstr. Weed Sci. Soc. Am.*, Vol. 36, pp. 2

Poss, K.M. (1992). Herbicidal triazolinones. US patent, US 5,125,958

Qasem, J.R. (2011). Herbicides Applications: Problems and Considerations, *Herbicides and Environment,* Kortekamp A. (Ed.), pp. 643-647. Available from: http://www.intechopen.com/articles/show/title/herbicides-applications-problems-and-considerations

Price, A. & Jessica Kelton, J. (2011). Weed Control in Conservation Agriculture, in: *Herbicides, Theory and Applications,* Soloneski S. & Larramendy M. L. (Ed.), pp. 3-13, InTech. Available from: http://www.intechopen.com/articles/show/title/weed-control-in-conservation-agriculture

Saito, K.; Sakaki, M.; Sato, R.; Nagano, E.; Hashimoto, S.; Oshio, H. & Kamoshita, K. (1993). New post-emergence herbicide S-23031; herbicidal activity and selectivity. *Abstr.18th Conf. Pest. Sci. Soc. Jpn.* p. 32

Scalla, R. & Matringe, M. (1994). Inhibitors of protoporphyrinogen oxidase as herbicides: Diphenyl ethers and related photobleaching molecules. *Weed Sci.*, Vol. 6, pp. 103-132, ISSN: 0043-1745

Thecodoridis, G. (1994). Herbicidal 2-[(4-heterocyclic-phenoxymethyl)phenoxy]-alkanoates US patent, US 5,344,812

Thecodoridis, G.; Bahr, J.T.; Hotzman, F.W.; Sehgel, S. & Suarez, D.P. (2000). New generation of protox-inhibiting herbicides. *Crop protection*, Vol. 19, pp. 533-535, ISSN: 0261-2194.

Theodoridis, G. (1997). Structure–activity relationships of herbicidal aryltriazolinones. *Pestici. Sci.*, Vol. 50, pp. 283-290, ISSN: 1096-9063

Theodoridis, G. (1989). Herbicidal aryl triazolinones. US patent, US 4,818,275

Theodoridis, G; Hotzman, F.W.; Scherer L.W.; Smith, B.A.; Tymonko, J.M. & Wyle, M.J. (1990). Synthesis and structure-activity relationships of 1-aryl-4-substituted-1,4-dihydro-5H-tetrazol-5-ones, a novel class of pre- and post-emergence herbicides. *Pestic. Sci.*, Vol.30, pp. 259-274, ISSN: 1096-9063

Vansaum, W.A. Bahr, J.T. Bourdouxhe, L.J. Gargantiel, F.J.; Hotzman, F.W., Shires, S.W.; Sladen, N. A.; Tutt, S.F. & Wilson, K.R. (1993). F84260- a new rapidly acting, low rate herbicide for the post-emergence selective control of broad-leaved weeds in cereals. *Proc. Brighton, Crop. Prot. Conf. Weeds* pp.19-22

Vencill, W.; Grey, T & Culpepper, S. (2011). Resistance of Weeds to Herbicides, *Herbicides and Environment,* Kortekamp A. (Ed.), pp.585-590, InTech, Available from:

http://www.intechopen.com/articles/show/title/resistance-of-weeds-to-herbicides

Vidrine, P.R.; Griffin, J.L.; Jordan, D.L. & Reynolds, D.B. (1996). Broadleaf weed control in soybean (Glycine, max) with sulfentrazone. *Weed Technol,* Vol. 10, pp.762-765, ISSN: 1550-2740

Wenger, J.; Winternitz, P. & Zeller, M.(1988). Heterocyclic compounds. PCT patent, WO 8810254

Witkowski, D.A. & Halling, B.P. (1988). Accumulation of photodynamic tetrapyrroles induced by acifluorfen-methyl. *Plant Physiol.* Vol. 87, pp. 632-637, ISSN: 1532-2548

Yoshida, R.; Sakaki, M.; Sato, R.; Haga, T.; Nagano, E.; Oshio, H.& Kamoshita, K. (1991). S-53482- a new N-pehenyl phthalimide herbicide. *Proc. Brighton Crop Prot. Conf. Weeds* pp. 69-75

Yoshimura, Y.; Ukai, S.; Hori, M.; Hirai, K.; Yano, T. & Ejiri, E. (1992). New oxazolidinedione herbicide; mode of action of paddy field herbicide KPP-314. *Abstr. 17th Conf. Pest. Sci. Soc. Jpn.* pp. 48

Zhang, L.; Tan, Y.; Wang, N.X.; Wu, Q.Y.; Xi, Z. & Yang, G.F. (2010). Design, syntheses and 3D-QSAR studies of novel N-phenyl pyrrolidin-2-ones and N-phenyl-1H-pyrrol-2-ones as protoporphyrinogen oxidase inhibitors. *Bioorg. Med. Chem.,* Vol. 18, pp. 7948-7956, ISSN: 0968-0896

6

Chemical Behaviour and Herbicidal Activity of Cyclohexanedione Oxime Herbicides

Pilar Sandín-España, Beatriz Sevilla-Morán,
José Luis Alonso-Prados and Inés Santín-Montanyá
Instituto Nacional de Investigación y Tecnología Agraria y Alimentaria (INIA)
Spain

1. Introduction

Benzoximate (I; Fig. 1) is an acaricide developed by Nippon Soda in 1971 (Iwataki, 1992). However, scientifics of the company observed that some benzohydroxamates showed weak herbicidal activity. After much synthetic developments, a new lead compound, an ethoxyimino dehydroacetic acid derivative (II; Fig. 1), showed a strong pre-emergence herbicidal activity against annual grass weed without any effects towards broadleaf plants. Further developmental research was performed on the cyclohexanedione skeleton to develop a new post-emergence herbicide. It was observed that the ethoximine group between the two keto groups was essential for the herbicidal activity (III; Fig. 1). Besides, when hetero atoms were introduced in the ring, the compounds showed high pre-emergence activity. When the ring was formed by carbons, so-called cyclohexane derivatives, the compounds showed high pre- and post-emergence activities. Therefore, the synthetic research was focused towards the substituents on the cyclohexanodione skeleton. The activity was higher when side chain substituents R_1 and R_2 (Fig. 1) were alkyl groups. As for the ring substituents, mono i-Pr and germinal dimethyl at the R_3 and $R_{3'}$ position and ciano and methoxycarbonyl groups at the R_4 position provided the maximum activity. This way, alloxydim–sodium was discovered and introduced in the market in 1978.

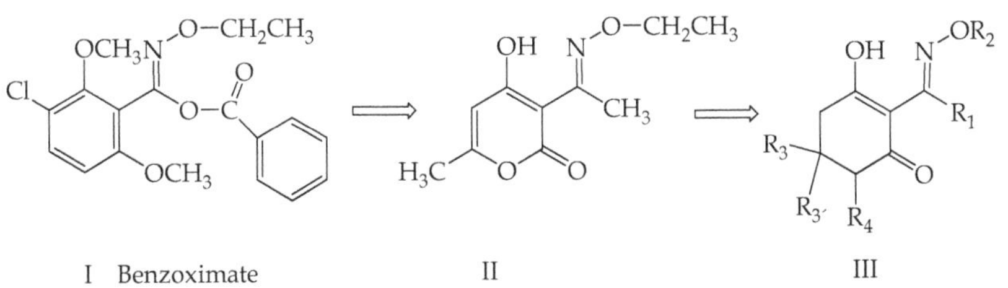

     I   Benzoximate   II   III

Fig. 1. Pathway of lead compounds towards the discovery of cyclohexanedione oxime herbicides.

However, though alloxydim-sodium showed a potent activity against annual grass weeds, it did not against perennial grass weeds. Therefore, the synthetic research was focused towards the introduction of different substituents on the cyclohexane ring, since the structure-activity pattern of the skeletal had been already identified. It was disclosed that substituents having a hetero atom such as chlorine, oxygen or sulfur increased the herbicidal activity. Some sulphur-containing cyclohexanodione derivatives showed very high activity against annual and perennial grass weed with post-emergence treatment. Thus, alkyl- and aryl- thioalkyl groups were introduced in position R_3 (III; Fig. 1) resulting in the discovery of other cyclohexanedione herbicides such as, sethoxydim, clethodim or cycloxydim (Fig. 2).

Thus sethoxydim, introduced in 1982, showed excellent herbicidal activity against various weed species in particular *Sorghum halepense*. Afterwards, clethodim, discovered by Chevron Chemical (Tomlin, 2006), presented almost the same herbicidal spectrum to sethoxydim but the application rates seems to be lower (Iwataki, 1992). Cycloxydim, discovered by BASF, showed a very broad spectrum with application rates similar to sethoxydim. Other herbicides like tralkoxydim have an herbicidal spectrum narrower than sethoxydim and clethodim. It has been used only for the control of annual winter grass weeds in wheat fields (Roberts, 1998). The last cyclohexanedione oxime herbicide was profoxydim (Fig. 2) developed by BASF, and first registered in 1998 for the control of grass weeds in rice.

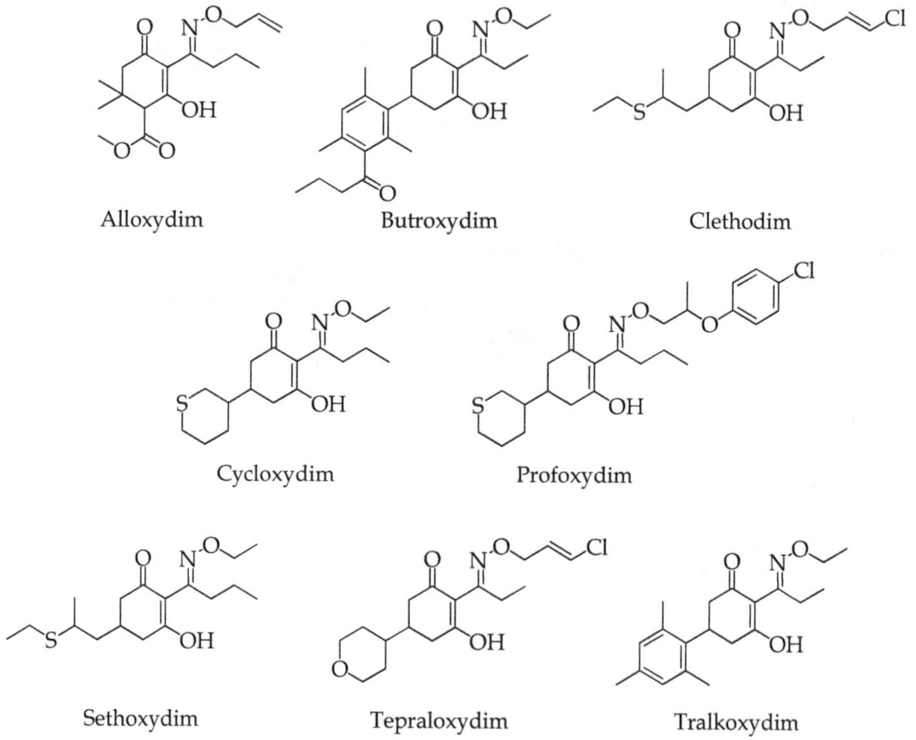

| Alloxydim | Butroxydim | Clethodim |

| Cycloxydim | Profoxydim |

| Sethoxydim | Tepraloxydim | Tralkoxydim |

Fig. 2. Chemical structures of cyclohexanedione oxime herbicides.

The basic structure of the cyclohexanedione oxime herbicides is shown in Fig. 3. These compounds show a keto-enol tautomerism, where the enolic forms (IV and VI) is generally predominant to the keto form (V) (Iwataki, 1992; Iwataki & Hirono, 1978). Thus, general formulas for this class of compounds are expressed as the enolic form. The term "dione" is used, however, as the general term for these compounds because of its simplicity.

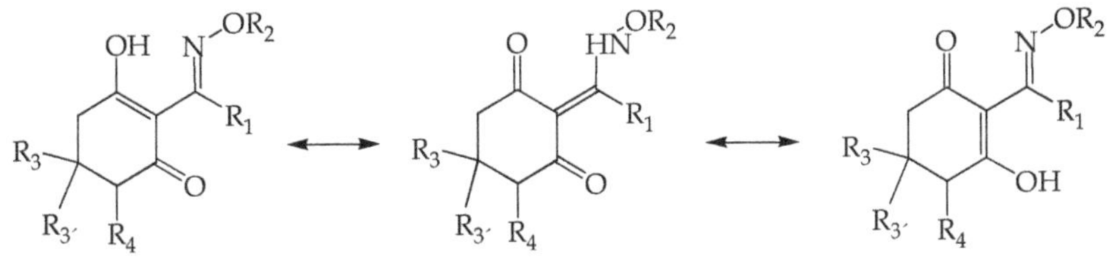

IV   V   VI

Fig. 3. Keto-enolic tautomerism of cyclohexanedione oxime herbicides.

As can be observed in Fig. 3, these herbicides present two isomers E and Z relating to the chloroallyloxy side chain, though the E form is more active and more stable than the Z form (McInnes et al., 1992; Sandín-España et al., 2003) and they are commercialized in this isomeric form.

Herbicides of this family are weak acids with pKa below 5. Thus, they can readily ionize, and if the pH increases above the pKa value of the herbicide, the ionized form will predominate. The protonated form of a herbicide will penetrate the plant cuticle more rapidly than the ionized form (Bukovac et al., 1971).

Their solubilities and partition properties are highly dependent on pH. They are easily decomposed by pH variations and by sunlight radiation (Roberts, 1998). These properties are highly relevant to their environmental fate.

Degradation is so rapid in aqueous media that in some cases it is questioned if herbicidal activity is maybe due to some degradation product (Iwataki, 1992). Due to the rapid degradation, most degradation products are common to most systems and in some cases it is not clear if a by-product is produced by chemical or biochemical degradation.

Therefore, it is of utmost importance to study the degradation routes of these herbicides to estimate the persistence of the residues of these compounds and to identify the factors that influence their behaviour in the environment.

2. Abiotic transformations, degradation pathways and degradation products

When an herbicide is introduced into the environment, it is subjected to different biotic and abiotic processes. Abiotic transformations may include chemical (mainly hydrolysis and thermolysis) and photochemical reactions. Complete mineralization of herbicides into inorganic constituents such as carbon dioxide, ammonia, water, mineral salts, and humic substances often occurs slowly in the environment. Different compounds, however, are formed before herbicides can be completely degraded. The organic compounds formed by the different transformation processes are referred to by several names such as degradates, by-products, etc. The most widely used are "transformation product" and "degradation product". "Metabolites" is a term usually inappropriately employed as it should be use only when the transformation product is a result of biological transformation. Therefore, in this chapter mainly dedicated to abiotic processes, we will refer to these compounds as degradation products, by-products or photoproducts if they are formed by transformations induce by sunlight.

In general, one or two transformations in the molecular structure of some herbicides are enough to modify its properties. In fact, the biological activity and/or environmental contamination attributed to the parent compound can be due to the degradation products.

The understanding of the total consequences for herbicide use is limited to the fact that most studies have focused on the parent compound and generally did not consider their transformation products (Somasundaram & Coats, 1991).

Historically, some of the most serious concerns about the safety of herbicides have raised from its transformation products than can cause detrimental side effects. Different studies confirm that many degradation products are more mobile and some others are more persistent than their respective parent compound (Boxall et al., 2004; Green & Young, 2006; Kolpin et al., 2004; Tixier et al., 2000).

Information about these degradation routes is necessary to estimate the persistence of these compounds and to identify the factors that influence their behaviour in the environment.

Nowadays, some herbicide transformation products are considered as "emerging pollutants" (Richardson, 2009; Rodríguez-Mozaz et al., 2007) as most of them have been presented in the environment for a long time, but their significant and presence are only now been elucidated and, therefore, they are generally no included in the legislation. Besides, there is still a lack of knowledge about long-term risks that the presence of these emerging pollutants may pose for organisms as well as for human health. Consequently, transformation products have become a new environmental problem and have awakened great concern among scientists in the last years (Richardson, 2006; Richardson, 2007).

Besides, herbicide degradates can either be less toxic or have similar or greater toxicity than their parent compounds (Belfroid et al., 1998; Tuxhorn et al., 1986). Thus, obtaining data on parent compounds and their primary degradation products is critical for understanding the fate of herbicides in the environment.

As mentioned before, the different abiotic/biotic processes that take place in the environment modified the physicochemical properties of the parent molecule.

Most of the oxidative reactions (hydroxylation, sulfoxidation, dealkylation...) and hydrolysis impart also some degree of increase polarity and hence water solubility to the molecule. Therefore, the new xenobiotics are more mobile in soil. Reduction reactions are characterized in environments with low oxygen concentrations, low pH and anaerobic microorganism. These reactions are less commonly observed and generally give rise to products with lower polarity.

Recent studies indicate that about one-third of the degradation products derived from a range of pesticides types have an organic carbon absorption coefficient (K_{oc}) of at least one order of magnitude lower than that of the corresponding parent compound. Thus, these transformation products may be more likely to be transported to surface and groundwater.

Boxall and co-workers (Boxall et al., 2004) showed that among different classes of pesticides and its transformation products, 41% of the transformation products were less toxic than parent compound and 39% had a similar toxicity to their parents, but 20% were above 3 times more toxic and 9% were above ten times more toxic than their parent compounds.

Oxidation reactions occur frequently in some soil and are an extremely important transformation pathway. S-containing herbicides, like some cyclohexanediones, are often rapidly oxidized to sulfoxide and afterwards more slowly to sulfones (Roberts, 1998). Sulfoxidation of herbicides can occur in soil and water mediated by chemical or biological reactions. This oxidation is so rapid and complete that sulfoxides are often the compounds found in soil shortly after application of the parent sulfide compound. Furthermore, in some cases, sulfoxides and sulfones are suspected to have pesticidal activity (Ankumah et al., 1995; Campbell & Penner, 1985; Tuxhorn et al., 1986).

2.1 Chemical degradation

Chemical degradation of organic compounds includes mainly hydrolysis and thermolysis reactions. With regard to hydrolysis, pH of water is responsible for the transformation of some pesticides in solution, especially in conjunction with extreme pH (García-Repetto et al., 1994).

Furthermore, organic compounds sensitive to pH give rise to its rapid degradation even with a slight variance of pH (Dannenberg & Pehkonen, 1998; Santos et al., 1998; Sanz-Asencio et al., 1997) and hence they have a low environmental persistence.

Alloxydim-sodium is a sodium salt of an acid, alloxydim (Fig. 4), having a pKa of 3.7, which is present as the monoanion and/or the acid in aqueous solution, and the possibility of many tautomeric forms should be considered to understand the transformations in the molecule when degradative reactions occur.

Alloxydim-sodium is neutralized to the sodium free compound (alloxydim) by the action of carbon dioxide in air or components in plants or in soil.

Studies of thermal degradation of alloxydim gave two oxazole derivatives (VII and VIII; Fig. 4) at 120 °C. The ratio was found to be 3 to 2 (VII:VIII) (Iwataki & Hirono, 1978). The mechanism of the formation of these oxazoles seems to take place when in certain tautomeric isomers, the cyclohexane ring of alloxydim is coplanar with the six membered ring which is formed by hydrogen bonding and the allyloxy group should be in the anti position against the cyclohexane group on the C=N bond. Therefore, the Beckmann rearrangement reaction occurs, which coincides with the intramolecular cyclization to form the oxazoles.

In the same way, when alloxydim is heated at 30, 40 and 50 °C in a dark incubator for 20 days gives the mixture of VII and VIII (Fig. 4). These three degradation products also appeared when alloxydim-sodium is applied in the leaves of plants (Hashimoto et al., 1979a; Koskinen et al., 1993).

The acid hydrolysis of alloxydim gives rise mainly to the butyrylamido derivatives and the imine salt. The alkaline hydrolysis forms the demethoxycarbonylated butyrylamido derivative (Iwataki & Hirono, 1978).

Chemical degradation of herbicides can also take place when the herbicide gets in contact with water that possesses substances that promote its degradation. In this sense, it is known that the presence of substances employed for the disinfection of water such as hypochlorite and chloramines degrade the herbicide to compounds more or less toxic than the active substance (Lykins et al., 1986; Magara et al., 1994; Reckhow & Singer, 1990).

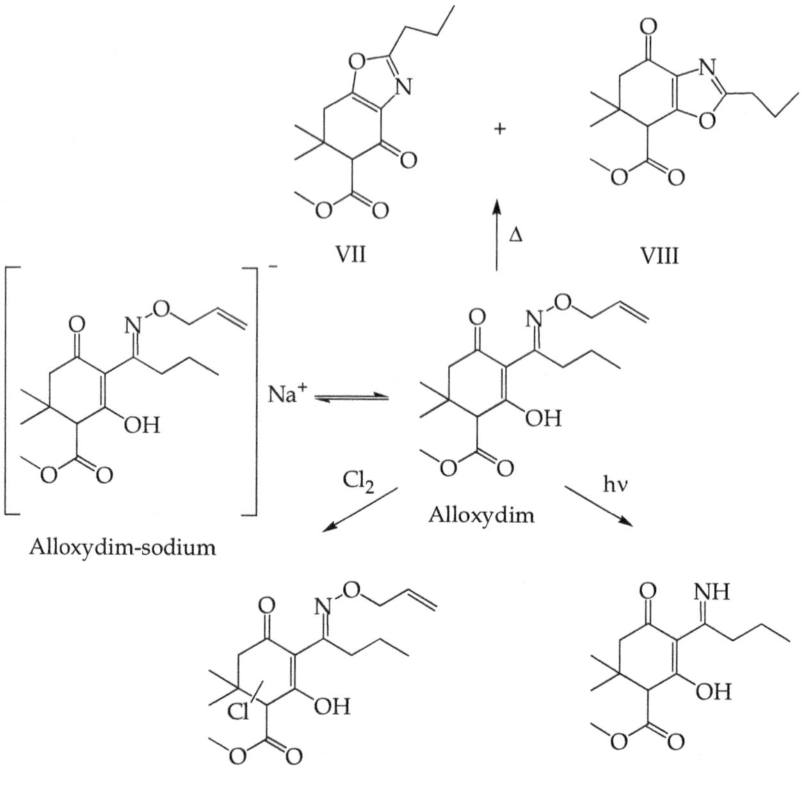

Fig. 4. Main degradation pathways of alloxydim-sodium.

Studies of alloxydim degradation in chlorinated water showed that the herbicide degrades very fast with half-lives less than one minute. As a result of the reaction one degradation product is formed that is isolated by solid phase extraction and identified by means of mass spectrometry as a chlorinated cyclohexanedione oxime compound IX (Fig. 4) (Sandín-España et al., 2005b).

Other cyclohexanedione oxime herbicide, clethodim is also an labile acid. Degradation rates significantly decreases from pH 7 to pH 5. After 20 hours, clethodim loss was 37% at pH 5 , 8% at pH 6 and 0% at pH 7 (Falb et al., 1990). Nine by-products were separated by liquid chromatography, though none of them was identified.

Falb et al., 1991 (Falb et al., 1991) also observed an increased in the degradation of clethodim when pH decreased. 19 peaks of degradation products more polar than parent clethodim molecule were separated by liquid chromatography.

As mentioned before, cyclohexanedione herbicides are marketed as the E-isomer at the oxime ether double bond (Fig. 5). It has been stated that some of them, like clethodim, may equilibrate with the Z-isomer in a polar medium such as water (Falb et al., 1990; Sandín-España et al., 2003). 4% of isomerization of clethodim to Z-isomer was observed when preparation of aqueous solution and 40% after two months (Sandín-España et al., 2005a). This isomerization has been also observed in herbicide tepraloxydim where equilibrium between both isomers was slowly attained in aqueous solution. It took about 7 days to reach equilibrium with a final ratio between isomers of 2:1 (Z:E) (Sandín-España et al., 2003). Other studies show that isomer Z also appears in acidic water samples of E-tepraloxydim after 72 h of storage at 4 °C (Sandín-España et al., 2002).

Degradation of clethodim in chlorinated water either with sodium hypochlorite or chloramines was very rapid with half-lives below 10 minutes (Sandín-España et al., 2005a). The main degradation processes was the oxidation to clethodim sulfoxide (Fig. 5). Experiments continued to follows degradation of the sulfoxide molecule. Its half-life was 4.4 seconds with hypochlorite and 9.3 hours with chloramines. Subsequent oxidative reaction of the sulfoxide generates the formation of clethodim sulfones (Fig. 5) and other minor products. The fast degradation of clethodim in chlorinated water (either with hypochlorite or chloramines) practically precludes any possible exposure of consumers to this compound when tap water is subjected to a chlorinated treatment. However, it is not possible to ensure complete destruction of the reaction product clethodim sulfoxide before the distribution point when chloramines are used for disinfection due to the slower degradation rate. Besides, whereas some minor degradation products remain unidentified, none of the major degradation products of clethodim contain more chlorine atoms than the parent compound which represents a positive aspect of this compound with respect to consumers' safety.

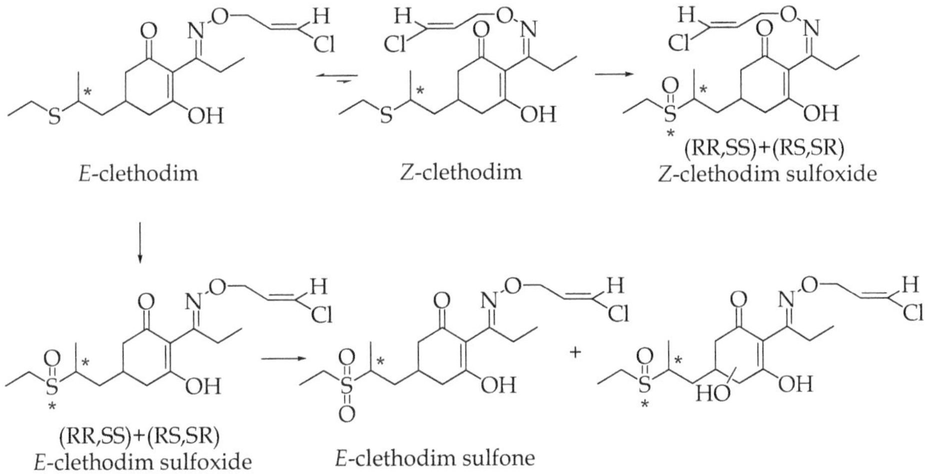

Fig. 5. Proposed degradation pathway of clethodim in chlorinated water.

Sethoxydim dissolves in water is found to be unstable at room temperature or when kept at -20° C; only 6 and 24% or the parent sethoxydim remained after 72 hours (Campbell & Penner, 1985). This herbicide also undergoes chemical decomposition at acid pH (Shoaf & Carlson, 1992; Smith & Hsiao, 1983) and in alkaline solution (Shoaf & Carlson, 1986).

Others cyclohexanedione herbicides such as profoxydim or tralkoxydim have also been reported to undergoes hydrolysis depending on the pH (Walter, 2001) of the solution and being easily degraded in aqueous solution (Sevilla-Morán et al., 2011; Srivastava & Gupta,

1994) and in chlorinated water (Sandín-España, 2004). However, scarce data exists in the open literature about degradation pathways or degradation products.

2.2 Photochemical degradation

Photochemical reactions are one of the major transformation processes affecting the fate of pesticides in the environment, especially in the aquatic compartment (Dimou et al., 2004; Neilson & Allard, 2008). Two ways of photodegradation reactions occur in sunlit natural water. In direct photolysis, organic compounds absorb light and as a consequence of that light absorption, undergo transformation. For this to occur in the water, the emission of the sun (290-800 nm) needs to fit the adsorption spectrum of the pesticide. In indirect photochemical reactions organic chemical are transformed by energy transfer from another excited species (e.g., components of natural organic matter) or by reaction with very reactive, short-lived species formed in the presence of light (e.g., hydroxyl radicals, single oxygen, ozone, peroxy radicals, etc.). Absorption of actinic radiation by nitrate and dissolved organic matter (DOM) leads to the formation of most of these species. Therefore, the composition of the aquatic media plays an important role on the phototransformation of pesticides in this compartment.

The hydroxyl radical, $OH^{\bullet}$, is one of the most reactive of the aforementioned reactive intermediates due to its non-selective and highly electrophilic nature.

Ideally solar radiation should be used in studies of environmental photochemistry, however, meteorological conditions in most countries and a slow degradation rate do not permit reproducible experimentation. In a first approach to study the photochemical behavior of organic compounds in different matrices, it is common to conduct the degradation under controlled conditions. Generally, the use of xenon arc lamp, with light above 290 nm (provided by a filter), is preferred as its spectral emission distribution are very close to the solar radiation spectrum (Marcheterre et al., 1988). It has been demonstrated that the use of different light sources under identical aqueous conditions can produce similar degradation products, with the only difference being in their kinetic of formation (Barceló et al., 1996). As for the experimental equipment, quartz glass is preferred instead of other glass material since it permits a greater transmission of radiation (Peñuela et al., 2000).

The composition of aquatic media also plays an important role in the phototransformation of pesticides. Various authors point out that particulate matter, such as sediment particles, and dissolved substances present in natural waters could be responsible for the different photolysis rates observed between natural and distilled water (Dimou et al., 2004; Schwarzenbach Rene et al., 2002; Tchaikovskaya et al., 2007). The most important light-absorbing species that may induce indirect photolytic transformation of organic pollutants in natural waters are the chromophores present in dissolved organic matter (DOM) where humic acids are important absorbing constituents of it and in a lesser extent, fulvic acids.

Diverse studies are available from literature where humic acids act enhancing (Sakkas et al., 2002a, 2002b; Santoro et al., 2000; Vialaton & Richard, 2002) or inhibiting (Bachman & Patterson, 1999; Dimou et al., 2004; Dimou et al., 2005; Elazzouzi et al., 1999; Sevilla-Morán et al., 2010a; Sevilla-Morán et al., 2008) the degradation of pesticides. In the first case, humic acids behave as a "sensitizer" where the excited states of humic acids can participate in a charge-transfer interaction with pesticides, or generate reactive intermediates, such as hydroxyl radicals, singlet oxygen, solvated electrons or hydrogen peroxide. In the second case, humic acids act as photon trap (optical filter effect), decreasing the photodegradation rate of pesticide.

Consequently, information about photodegradation of pesticides is necessary to estimate the persistence and to identify the factors that influence their behavior in the environment. Furthermore, it is important to investigate what these compounds degrade into, its persistence of by-products relative to the parent compounds, and whether the degradation products retain the activity of the active substance to cause a toxicological effect on non-target organisms in aqueous systems.

Studies show that alloxydim is degraded on the leaf surface by photochemical reactions. After two days, the deallyloxylated compound X (Fig. 4) is the main degradation product identified. This by-product is formed by photoreduction. Other minor degradation products identified, that accounted less than 0.5% of the applied radioactivity, were two isomeric oxazoles and a demetoxycarbonylated compound (Hashimoto et al., 1979b; Soeda et al., 1979).

When photodegradation of alloxydim was studied in sterilized soil (Ono et al., 1984), only 22% of the ^{14}C-alloxydim applied was remained. The main by-products identified by thin-layer chromatography and mass spectrometry were again the deallyoxylated compound (X) and the two oxazole isomers (VII, VIII) (Fig. 4). The demethoxycarbonylated compound was formed in a minor extent (Ono et al., 1984).

Photochemical transformation of a methanol solution of ^{14}C-alloxydim on silica gel plate, irradiated with UV light also gives by-products VII, VIII and X (Soeda et al., 1979).

Sevilla-Morán et al., studied the photodegradation of alloxydim-sodium under simulated solar irradiation using a xenon arc lamp (Sevilla-Morán et al., 2008). This light source fits the solar radiation spectrum best over the whole range of spectral emission (Marcheterre et al., 1988). Indirect photolysis under the presence of various concentrations of humic acids, nitrate and iron ions was also investigated. Results show that degradation rate in direct photolysis was higher as the radiation intensities increased. Irradiation of aqueous solutions of alloxydim containing different concentrations of humic acids (1-20 mg L^{-1}) show that increasing concentrations of humic acids, decrease photolysis rate of this herbicide, indicating that absorbed most of the photons emitted, thereby slowing down direct photochemical reaction of alloxydim. The presence of nitrate ions had no effect on the degradation rate. On the contrary, iron ions accelerate the rate of photolysis of alloxydim possibly due to the formation of a complex and later undergo a direct photolysis (Boule, 1999; Park & Choi, 2003). Simultaneously to the irradiation experiments, control experiments in absence of radiation were performed in order to discard other type of dark reactions (hydrolysis, thermolysis, ...). Quantitative recoveries of clethodim during the entire exposure period to simulated solar irradiation enable to ignore other transformation processes that are not initiated by radiation.

As for the identification of transformation products, HPLC-ESI-QTOF-MS technique was employed. QTOF provides elevated resolution and sensitivity, high mass accuracy for both parent and fragment ions in combination with the possibility of performing MS/MS acquisitions obtaining more structural information (Aguera et al., 2005; Ibáñez et al., 2004). In this way, QTOF allows the assignments of a highly probable empirical formula for unknown compounds. Two main degradation pathways were identified. The main reaction was the photoreduction of the N-O bond dissociation of allyloxyamino moiety to give the imine (X; Fig. 4). The second reaction was the isomerization of the oxime moiety to give the Z-isomer (Sevilla-Morán et al., 2008).

In the same way, under UV light technical clethodim was greatly accelerated as compared to dark conditions. The degradation half-lives were 2.5, 2.6 and 3.2 hours at pH 5, 6 and 7 (Falb et al., 1990). The HPLC system separated 13 photoproducts though none of them were identified. The addition of adjuvants to clethodim increased the rate of photodegradation with UV radiation by 2 to 7 fold. The rate of degradation under sunlight was increased with the addition of adjuvant by 7 to 27 fold over the control (Falb et al., 1990).

Falb and co-workers, (Falb et al., 1991) also observed that clethodim undergoes degradation when exposed to UV light and developed a method for the separation of 31 degradation products, but none of them were characterized.

It is recommended that spraying at late evening or night may improve cyclohexanedione oxime herbicides efficacy due to a reduction in the amount of UV light present (McMullan, 1996). Applying cyclohexanedione herbicides at times when the UV light is lowest, such as late evening or at night, will maximize weed control with these herbicides.

Cyclohexanedione herbicide efficacy was affected by spray solution pH. This was probably due to the predomination of the ionized form of the cyclohexanedione molecule at high spray solution pH and the possible formation of cyclohexanedione sodium salts (Nalewaja et al., 1994).

Experiments on the efficacy of clethodim and tralkoxydim (McMullan, 1996) showed that filtering UV light 4 hours after treatment improved the efficacy between 13 and 55%. These results are in agreement with results published by McInnes et al. with sethoxydim (McInnes et al., 1992).

Indirect photodegradation of clethodim in the presence of humic acid, nitrate ions and iron ions in aqueous solution has been also studied (Sevilla-Morán et al., 2010a). The presence of humic acid increased the half-life of herbicide photolysis compared to direct photolysis $[t_{1/2([HA]=1\ mg/L)}$ = 44,2 min $vs.$ $t_{1/2ultrapure\ water}$ = 28,9 min]. This retarding effect, as in alloxydim, indicates that these substances can act as an "optical filter" absorbing most of the photons emitted and thereby slowing the direct photochemical reaction of clethodim. Nitrate ions had no effect on the photodegradation of clethodim and the presence of iron ions increase the rate of photolysis up to 6 times when 20 mg L^{-1} of Fe(III) ions were present in the solution. In these experiments up to nine different by-products were observed all of them more polar than clethodim (Sevilla-Morán et al., 2010a). Identification was performed by using a QTOF mass spectrometer. A detailed study based on the exact mass measurements and fragmentation pattern makes possible for the first time to elucidate the structures of the nine photodegradation by-products. Figure 6 shows the proposed photodegradation pathway of clethodim in aqueous solution. The by-products identified were the following; Z-isomer of clethodim at the oxime ether double bond. The Z-isomer is much more polar that clethodim (26 $vs.$ 41 minutes). This could be due to an internal hydrogen bond formed between the oxime oxygen and the hydroxyl group of the cyclohexane ring. Several authors stated that some E-isomer of cyclohexanedione oxime herbicides may equilibrate with the Z-isomer in polar solvents (Falb et al., 1990; Sandín-

España et al., 2002) or in chlorinated water (Sandín-España et al., 2005a). Besides, it has been reported that isomerization can be induced by light and temperature (Curtin et al., 1966; Sevilla-Morán et al., 2008).

The oxidation of the sulphur atom of the molecule generates two pairs of enantiomers (RR+SS and RS+SR) and diastereomers are chromatographically separated in two peaks, containing a pair of enantiomers each. In the same way, Z-isomer of clethodim gave rise to the corresponding pairs of enantiomers of Z-clethodim sulfoxide. As in alloxydim, photoreductive reaction of E- and Z-clethodim forms the corresponding imine. Oxidation of imine gave rise to two pair of enantiomers of clethodim imine sulfoxides. The oxidative cleavage of the C-S bond of clethodim imine gave rise to clethodim imine ketone.

Various authors have also studied the photodegradation of sethoxydim (Campbell & Penner, 1985; Sevilla-Morán, 2010; Sevilla-Morán et al., 2010b; Shoaf & Carlson, 1986; Shoaf & Carlson, 1992). The herbicide was completely lost within seconds in aqueous media either in incandescent or UV light at pH 3.3 and 6.0 and methanolic solutions of the herbicide were transformed more than 50 % after 10 min of exposition to UV light (Shoaf & Carlson, 1992) (Shoaf & Carlson, 1992). Upon 1 hour of UV irradiation sethoxydim sulfone was identified as one of the degradation products formed during the experiment. Campbell and Penner suggested the rapid degradation of sethoxydim in water and organic solvents (Campbell & Penner, 1985). These authors exposed aqueous solutions of sethoxydim to artificial light and observed that only 2 % remained after 3 h. In the same way, they also observed a rapid photodegradation on glass disks of sethoxydim dissolved in n-hexane (81 % of herbicide was transformed after 1 h). In both systems, 6 major products were detected. Five of these products were transitory and only one appeared to be the single end product. One of these compounds was isolated and identified by mass spectrometry as desethoxy-sethoxydim.

Evidence of rapid transformation of sethoxydim expose to light suggested that herbicidal activity resulted from the more stable transformation products. Five of these products were applied to barnyardgrass to determine phytotoxicity. It was found that two degradation products had significant activity as did sethoxydim and three of them showed no herbicidal activity. Though none of these isolated by-products were quantified as relative herbicidal potencies could not be determined, this is relevant information about phytotoxicity of degradation products.

Under field conditions, were sethoxydim would be applicated by spraying water-oil emulsion during daylight, it is likely that sethoxydim transformation would be rapid and levels of degradation products would be present. These findings suggest that some of the degradation products are actually the herbicidal agent (Campbell & Penner, 1985; Shoaf & Carlson, 1992).

In order to obtain results close to real field conditions different experiments have been carried out to study the photodegradation of sethoxydim in natural waters (mineral, well and river) and under natural sunlight. The degradation rates in natural waters were lower than in ultrapure water and degradation under natural sunlight also decrease the degradation rates. These results indicate that under real environmental conditions photodegradation of sethoxydim is retarded compare to laboratory studies. Besides, significant differences among types of water suggest that degradation of sethoxydim has a strong dependence on the composition of water sample. Figure 7 shows the photo-

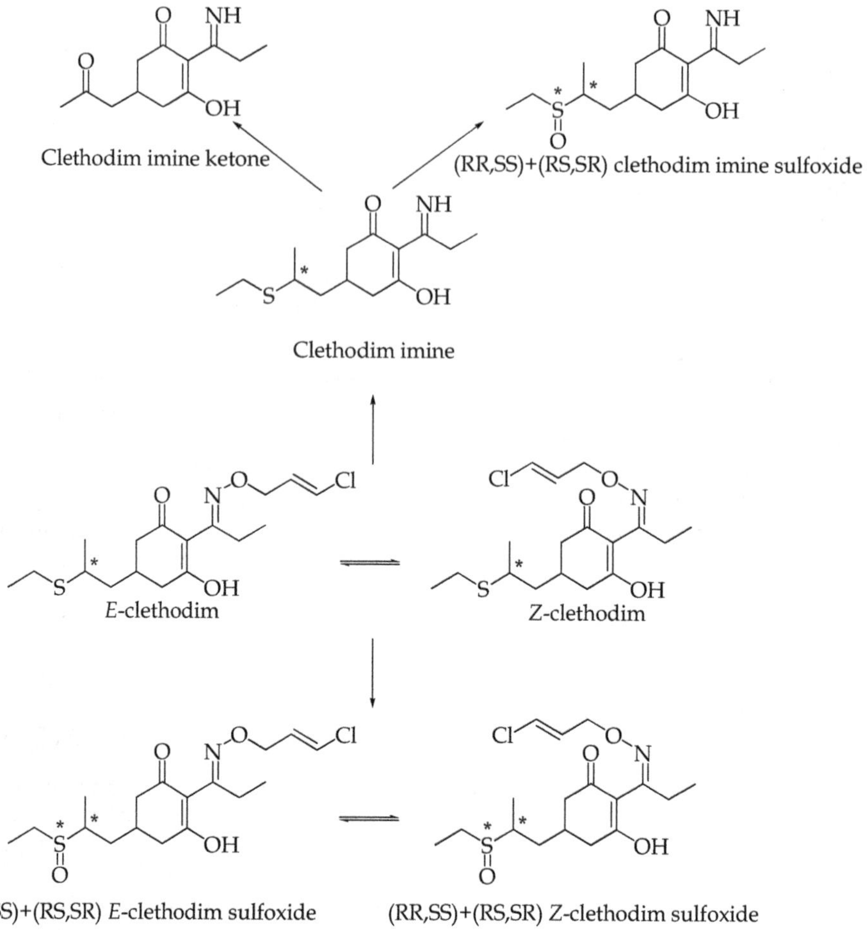

Fig. 6. Proposed photodegradation pathway of clethodim in aqueous solution.

degradation curves of sethoxydim in natural waters under solar irradiation. The photolysis rate decreases in the following order, river<well≈mineral<ultrapure water. Photodegradation of sethoxydim-lithium in natural water was approximately 5 times slower than in ultrapure water showing a half-live of 436.9 ± 0.8 min for river water and 82.1 ± 0.7 min for ultrapure water. These significant differences observed in the three water samples, under both types of irradiation, indicated that the degradation of sethoxydim-lithium depends to a large extent on the composition of aqueous matrix. Natural water contains some substance/s that induces a retardant effect of the photodegradation of sethoxydim-lithium. Thus, this difference observed (2-5 folds) could be attributed to the presence of increasing concentrations of TOC (Total Organic Carbon) in the natural waters, where river water has the highest concentration of TOC (2.865 mg L^{-1}) and ultrapure water has the lowest (0.005 mg L^{-1}).

The photodegradation of sethoxydim-lithium in all water types under simulated light was quicker than under natural sunlight (e.g. $t_{1/2(river\ water-natural\ sunlight)}$=436.9 ± 0.8 min $vs.t_{1/2\ (river}$

$_{water)}$ 135.5 ± 0.3), which is well correlated with the lower intensity of natural sunlight (460 W m^{-2} at the middle of the day $vs.$ 750 W m^{-2} of xenon lamp). These results show a great influence of the intensity radiation on the degradation of sethoxydim-lithium and they are in accordance with other authors where the degradation of active substance is dependent on the irradiation energy (Dimou et al., 2005; Sakkas et al., 2002a).

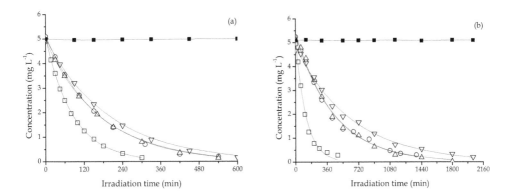

Fig. 7. Photodegradation of sethoxydim-lithium in various types of water under simulated light (a) and natural sunlight (b): (□) ultrapure water, (○) mineral water, (Δ) well water, (∇) river water, (■) dark experiment.

3. Bioassay methods to detect herbicide phytotoxicity

Weeds are a continuous problem in agriculture. The success of modern agricultural practices is due in part to the discovery and adoption of chemicals for weed control. The introduction of selective herbicides has greatly facilitated farmers' work by suppressing the need for manual weeding. Indeed, the tremendous increase in crop yields associated with the "green" revolution would not have been achieved without the contribution of these synthetic compounds. The abundance of high quality food in developed nations has eliminated concerns about access to food in these countries. However, concerns over the potential impact of pesticides on the environment have now arisen and as consequence, more pressing and more stringent pesticide registration procedures have been introduced.

The basis for much of the work done in weed control is the testing of the response to the herbicides. Bioassay methods have been developed to determine the residue level of many herbicides in soil and water (Fig. 8). There are different types of bioassays, depending on the species, the type of herbicide used, its mode of action, substrate and other environmental conditions, as well as the measured parameter. Their use offer several advantages such as the detection of very low phytotoxic residues and the detection of its bioavailability. Therefore, bioassays can be used to complement the analytical chemical methods and are useful tools to screen herbicide phytotoxicity and provide information about the phytotoxicity of herbicide residue in the soil at sowing time. The sensitivity, low cost and reproducibility of bioassays fulfill the criteria for a good technique.

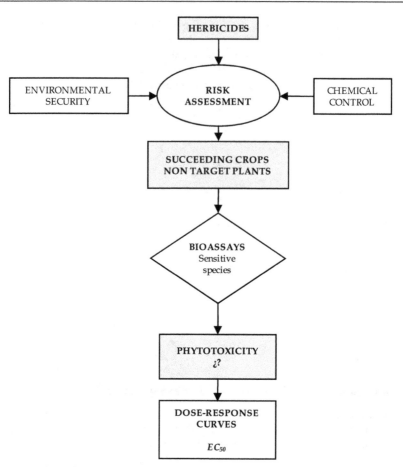

Fig. 8. Process of conducting bioassays in testing response of herbicides.

The classical bioassay, often used to quantify the amount of herbicide in soil, employs a single "standard" dose-response curve. This standard curve shows the plant response to different herbicide concentrations and report information of different concepts related to herbicide efficacy, such as selectivity, tolerance and resistance. These methods are of the utmost importance in studies of crop selectivity, herbicide resistance development and herbicide resistant weeds detection.

A typical dose-response curve is sigmoid in shape. One example of such a curve is the log-logistic curve (Seefeldt et al., 1994). The mathematical expression relating the response Y to de dose X is the following;

$$Y = C + \frac{D-C}{1+\exp\left\{b \cdot \left[\ln(X+1) - \ln(EC_{50}+1)\right]\right\}}$$

where C is the lower limit, D is the upper limit, b is the slope, and EC_{50} is the dose giving 50% response. The log-logistic is the most common model used in bioassays to describe dose-response relations. Other relevant sigmoid curves might be the Gompertz (Ritz et al., 2006), are used sometimes, for instance, in cases where a log-logistic model did not fit well to the data. Also, the toxicity factor used in the risk assessment is usually the EC_{50} (concentration required to give 50% reduction of the plant growth with respect to the control).

The consequences of herbicide introduction into the environment could not be limited to active ingredient. It is very important to know whether any phytotoxic effects detected due to active substance applied or to some of its metabolites or degradation products. In this case, it is necessary to know the route and rate of degradation not only for the active substance but also for the products of degradation (Berenzen et al., 2005).

Besides, the herbicide formulations contain many other compounds called adjuvants. Adjuvants are defined as "an ingredient in the pesticide prescription, which aids or modifies the action of the principal ingredient" (Foy, 1993). These products are added to increase the effectiveness (bioavailability) of the formulation by enhancing the solubility, or the compatibility of the active ingredients. Other functions can be to improve adsorption, penetration and translocation of the active ingredients into the target, increase rain fastness, and alter selectivity of the active ingredient toward different plants (Foy, 1993; Foy & Witt, 1992; Li & Foy, 1999). Furthermore, many papers concerning effect studies of pesticides have compared effectiveness of different adjuvants (Foy, 1993). Only very few papers discuss the environmental toxicity and risk of adjuvants; Chow and co-workers have previously noted this lack of information on the effects of adjuvants on the environment (Chow & MacGregor, 1983). Especially, information on both simple toxicological properties of adjuvants and their influence in complex ecosystems is required. Actually, the ecotoxicological aspects of the use of surfactants in cleaning products mostly for household have been published (Junghans et al., 2006; Krogh et al., 2003).

The potential problems that residues of these products may present, as the effect on non-target species or in successive crops, do not provide information needed to relate these effects with the chemical nature of the residue. Therefore, it is also necessary to study the nature of this residue by conventional analytical methods to identify potential causes of environmental problems.

3.1 Herbicidal activity of cyclohexanedione oxime herbicides

The graminicides are a group of commercially important selective herbicides (Harwood, 1999). One of major chemical classes of post emergence herbicides belongs to the cyclohexanedione class. This herbicides´ group is effective against a wide range of annual and perennial grasses (monocotyledonous) in a large variety of broad-leaved crop plants (dicotyledonous). Their biochemical target site is the enzyme acetyl coenzyme-A carboxylase (ACCase) (Burton et al., 1991; Rendina et al., 1990; Secor & Cseke, 1988), which catalyzes the first step in fatty acid biosynthesis. They are also known as group A herbicides or group 1 herbicides (Park & Mallory-Smith, 2004; Price et al., 2003). Since their introduction in the late 1970s, the ACCase-inhibiting herbicides have been widely used worldwide to control a number of grass weed species (Devine & Shimabukuro, 1994). Ciclohexanodione herbicides are also a family used at low-dose rate as they are biologically active at very low concentration (0.2-0.5 kg a.i. ha[-1]). It is currently being registered in

Europe, for weed control in broad-leaved crops, mainly in sugarbeet (*Beta vulgaris* L.), soya bean (*Glycine max* L. Merr.), and pea (*Pisum sativum* L.) crops, although its use is recommended in other broad-leaved crops such as rape (*Brassica napus* L.), potato (*Solanum tuberosum* L.) and beans (*Phaseolus vulgaris* L. and *Vicia faba* L.) (Sandín-España et al., 2003).

Cyclohexanedione oxime herbicides cause a rapid cessation of growth followed by destruction of shoot meristems in susceptible species. Studies on the uptake, translocation, and metabolic fate in tolerant and susceptible plants have shown that these herbicides inhibited *de novo* fatty acid biosynthesis in isolated chloroplasts, cell cultures, or leaves of susceptible grasses such as corn, wheat, and wild oats but not in tolerant broad-leaved plants such as soybean, spinach, and sugar beet (Burgstahler & Lichtenthaler, 1984). Injury symptoms tend to develop rather slowly in sensitive plants treated with cyclohexanedione herbicides. Growth (leaf elongation) stops within 24-48 h after herbicide application. Chlorosis is first observed on the youngest tissue, usually the emerging leaves. This reflects the fact that the initial phytotoxicity occurs primarily at the apical meristem, the major site of cell division and *de novo* fatty acid synthesis in these plants. In fact, 48-72 h after treatment the youngest emerged leaf can be quite easily separated from the rest of the plant by gently pulling it upwards; again, this reflects the tissue damage at the meristem. Chlorosis then spreads slowly through the rest of the plant, although it may take 7-10 days for the entire plant to be affected. Phloem translocation of these herbicides through the plant is limited, resulting in relatively small amounts reaching the roots. For this reason, these herbicides provide excellent control of perennial grass weeds. However, uncertain conditions some control of perennials can be achieved. No injury symptoms appear on dicotyledonous crops or weeds treated at typical use rates. Physiological injury can occur in cereal crops under certain conditions (e.g., low temperature at time of application), presumably due to reduced rates of herbicides detoxification. However, most plants recover from this temporary injury within 7-10 days (Walker et al., 1988).

The cyclohexanedione oxime class has been developed at low doses to reduce the adverse effects of the use of some herbicides and fulfill environmental requirements, water and soil pollution set by international legislation. However, due to high phytotoxicity, small amounts of residual herbicide in soil may affect sensitive succeeding crops. Also their polar character makes them easily leach to groundwater and potentially contaminate at levels above 0.1 µg L^{-1} (Sandín-España et al., 2003). In this context, studies about mobility, degradation and persistence in soil and water were performed with a variety of analytical techniques like gas and liquid chromatography, mass spectroscopy, photodegradation studies, studies with ^{14}C, immunoassays, etc. Most studies have been made in water and soil; occasionally there are some bioassays in microalgae (Santín-Montanya et al., 2007). The last results obtained confirm that could be susceptible specie capable to detect the presence of some herbicides (Table 1 & Fig. 9).

Herbicides	Log-logistic curve doses-response	R^2 (%)
Alloxydim	y= 0,19+((0,63-0,19)/(1+exp(1,74*(Ln(Dose+1)-Ln(285,22+1)))))	78,03
Sethoxydim	y=-0,0022+((0,66+0,0022)/(1+exp(0,62*(Ln(Dose+1)-Ln(110,33+1)))))	83,76
Metamitron	y=0,11+((1,08-0,11)/(1+exp(0,95*(Ln(Dose+1)-Ln(0,128+1)))))	98,14
Clopyralid	Not adjusted to regression equation	

Table 1. Regression equations by Seefeldt model that describe the relationships between increased doses of herbicides and growth of *Dunaliella primolecta* (unpublished data).

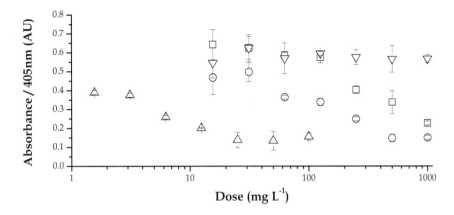

Fig. 9. Response of microalgae *Dunaliella primolecta* growth in presence of different concentrations of alloxydim (□), sethoxydim (○), metamitron (△) and clopyralid (▽) (unpublished data).

Other bioassays have been developed to detect phytotoxic residues of herbicide sethoxydim (Hsiao & Smith, 1983). In our group, initial attempts to obtain a practical hydroponic bioassay that allowed us to quantify tepraloxydim were frustrated due to the lack of repeatability and random results. Therefore, an investigation was carried out to determine the fate of tepraloxydim under bioassay conditions in order to clarify the reason for poor bioassay repeatability. The presence of residual chlorine in water was identified as the key factor on the repeatability of the bioassay.

Finally, an extensive research was conducted to develop and optimize a bioassay based on the high sensitivity of wheat (*Triticum aestivum* L.) to tepraloxydim in hydroponic culture using chlorine free mineral water (Sandín-España et al., 2003). Afterwards, similar studies were carried out with tralkoxydim (Table 2&3).

Dose (μg L^{-1})	Tepraloxydim		Tralkoxydim	
	Bioassay 1	Bioassay 2	Bioassay 1	Bioassay 2
0	11.9 a	12.1 a	9.5 a	11.9 a
2	10.2 b	10.7 b	8.2 b	11.9 a
4	6.7 c	5.8 c	7.7 b	9.3 b
8	3.5 d	3.2 d	3.9 c	3.9 c
16	1.5 e	1.6 e	1.4 d	1.2 d
32	0.7 e	0.7 e	0.8 d	0.6 d

Table 2. Mean values of root growth wheat treated with herbicides 7 days after treatment. Different letters after value, for each column, indicate differences at p<0.05.

	Tepraloxydim		Tralkoxydim	
	Bioassay 1	Bioassay 2	Bioassay 1	Bioassay 2
EC_{50} (µg L[-1])	4,6	3,8	6,8	7,0
R^2 (%)	99,8	97,2	98,4	99,6

Table 3. EC_{50} parameter of by Seefeldt model to two ciclohexanodione oxime herbicides in hydroponic culture of wheat.

It has been demonstrated that water chlorination with disinfection purposes degrades completely any possible residue of herbicide clethodim (Sandín-España et al., 2005a). This degradation is very rapid, giving rise to different degradation products. In this sense, we have studied the phytotoxicity of alloxydim and its main metabolite with hydroponic bioassays on wheat (Sandín-España et al., 2005b).

A chlorinated degradation product (IX; Fig. 4) of alloxydim was the main product obtained in its degradation with chlorine, one of the most common disinfectant agents employed in water treatment. Results showed that after seven days of treatment the most sensitive biological parameter for alloxydim was root length, causing in the root growth of plants a 40% of significative reduction at the dose of 0.3 mg L[-1] and 94% of reduction at the highest dose. However, the effect of metabolite on root growth only occurred at the highest metabolite dose (10 mg L[-1]), causing a 32% of reduction in root growth. Root system control presented normal growth (main tap root plus secondary roots), while those from injured plants were increasingly deformed (main tap root twisted and lack of secondary roots). Root growth was increasingly affected with doses from 0.1 mg L[-1] to the highest dose (Table 4).

The foregoing results suggested that the use of low dose herbicides can produce damage on succeeding crops, neighbouring crops and on non-target plants. Overall, there is no one species or endpoint that is consistently the most sensitive for all species or all chemicals in all soils, and differences in bioavailability among compounds may confound comparison of test results (Clark et al., 1993). Therefore, bioassays can provide additional information, with acceptable reproducibility (Ritz et al., 2006) on herbicide uptake and translocation (Horowitz, 1980; Reineke et al., 2002).

Dose (mg L[-1])	RG with Alloxydim (cm)		RG with Metabolite (cm)	
0	15.20 a		12.60 a	
0.1	13.67 a		13.15 a	
0.2	11.65 a		13.07 a	
0.3	9.04 b	↓ 40 %	13.59 a	
0.4	4.93 c		13.30 a	
0.5	2.64 d		13.35 a	
0.7	1.58 e		12.94 a	
1	0.90 e	↓ 94 %	12.23 a	
5			12.60 a	
10			8.55 b	↓ 32 %

Table 4. Response of root growth (RG) of wheat plants to different doses of alloxydim and its metabolite 7 days after treatment (means values of root growth). Different letters after value, for each column, indicate differences at p<0.05.

3.2 Occurrence of weed resistance to cyclohexanedione oxime herbicides

The reliance on herbicides for weed control has resulted in shifts in the weed flora and, more importantly, in the selection of herbicide-resistant weed populations. This is particularly true for herbicides with a single target, such as herbicides inhibiting acetyl coenzyme-A carboxylase (Devine & Shimabukuro, 1994).

Since their introduction to world agriculture in the 1980s, cyclohexanedione oxime herbicides have been widely used to control a variety of grass weeds. As a consequence, they rapidly selected, and are still selecting, resistant plants within grass weed species.

The resistance to acetyl coenzyme-A carboxylase-inhibiting herbicides was reviewed in detail (Devine & Shimabukuro, 1994), the number of grass weed species in which resistant plants have been reported increased from 9 to 34 (Delye et al., 2005) and resistance to ACCase-inhibiting herbicides has been reported in 26 countries. The estimates of cultivated land surfaces concerned by this resistance vary between 3 to 4.6 million hectares (Delye et al., 2005).

One of the best-studied weeds is black-grass (*Alopecurus myosuroides* Huds.), a major grass weed in winter crops in Europe. Similar findings were obtained in annual ryegrass (*Lolium rigidum* Gaud.) (Delye et al., 2003a), green foxtail (*Setaria viridis* L. Beauv.), wild oat (*Avena fatua* L.) and winter wild oat (*Avena sterilis* ssp. *Ludoviciana* Malzew.) (Christoffers & Kandikonda, 2006; Delye et al., 2003b; Delye et al., 2005; Shukla et al., 2004; Zagnitko et al., 2001), slender foxtail (*Eleusine indica* L. Gaertn.), barnyardgrass (*Echinochloa colona* L. Link) and little canarygrass (*Phalaris minor* Retz.).

Species	Location	Chemical class*
Avena fatua	Canada, USA, Australia, UK	AOPP, CHD
Avena sterilis	Australia, UK	AOPP
Alopecurus myosuroides	UK, Spain, Germany, France	AOPP
Digitaria ischaemum	USA	AOPP
Digitaria sanguinalis	USA	AOPP
Echinochloa colona	Costa Rica	AOPP
Eleusine indica	Malaysa	AOPP, CHD
Festuca rubra	USA	CHD
Lolium rigidum	Australia, Spain	AOPP
Lolium multiflorum	UK, USA	AOPP
Phalaris minor	Israel	AOPP
Setria faberi	USA	AOPP, CHD
Setaria viridis	Canada	CHD
Sorghum halepense	USA	AOPP

Table 5. Weed species exhibiting resistance to ACCase-inhibiting herbicides. Resistance has been conferred to one or more biotypes or accessions of the above weed species by reduced ACCase sensitivity. * AOPP: Aryloxyphenoxypropionate, CHD: Cyclohexanedione.

The occurrence of resistance in grass weeds, *Setaria faberi* Herrm. and *Digitaria sanguinalis* L.) Scop., to cyclohexanedione herbicides has been confirmed also in the United States (Stoltenberg & Wiederholt, 1995; Wiederholt & Stoltenberg, 1995). Sethoxydim had been the only ACCase-inhibiting herbicides applied to fields in which resistant plants were identified. In whole-plant dose–response experiments, resistant *S. faberi* was 134-fold resistant to sethoxydim (Stoltenberg & Wiederholt, 1995), but showed low levels of resistance to the cyclohexanedione herbicide clethodim. Similarly, Table 5 summarizes the weed species that are known to have developed resistance to ACCase-inhibiting herbicides around the world (Delye et al., 2005; Hatzios, 2001).

The development of resistance to ACCase-inhibiting herbicides in several grass weeds is an increasing problem in several parts of the world. Resistance to these herbicides can arise easily following selection pressure with cyclohexanedione herbicides for six to ten years. The judicious use of ACCase-inhibiting herbicides in combination with herbicides from other classes and methods of non-chemical weed control will be important for prolonging the usefulness of the cyclohexanedione herbicides.

4. Conclusion

The continuous use of plant protection products has led to the contamination of different environmental compartments, such as water, soil and air, being water contamination of great concern due to the risks for human consumption.

The synthetic research of herbicides in the last decades has shifted from long-life compounds to less persistent and more polar compounds, in order to avoid their accumulation in the environment. However, the low persistence of these compounds does not imply their completely mineralization but they are going to degrade to smaller molecules with different physicochemical properties than the active substances. In fact, it has been demonstrated that some of their transformation products are more mobile, persistent and/or more toxic than the parent molecule. Therefore, the knowledge of the fate of herbicides in the environment is underestimated if we do not take into account their transformation products whose behaviour and agro-environmental fate is in many cases unknown.

Besides, the lack of data on the phytotoxic effects of herbicide residues has been highlighted. In this sense, it is necessary to study and develop simple methods for evaluating the environmental impact of these products based on hard scientific data. Furthermore, it is also important not only study the phytotoxicity of the herbicide by means of bioassays, but also of their degradation by-products. From these studies should be able to derive recommendations for agricultural practices for the use of these products are environmentally friendly in general and in particular the agricultural environment capable of guaranteeing the future productivity of farms in the context of sustainable agriculture.

Cyclohexanedione oxime herbicides have been developed for the post-emergence control of grasses in dicotyledonous agricultural crops. These herbicides are unstable in aqueous solution and are very sensitive to pH and sunlight. These properties are highly relevant to their environmental fate. Degradation is so rapid that degradation products apparently could contribute to the activity of the parent molecule. Until now, some of these degradation products and the degradation pathways have been identified. However, the fate, significance and phytotoxicity of their degradation products is not fully known and future research is still need to attain a complete understanding of the fate of cyclohexanedione herbicides in the environment.

5. Acknowledgment

This work was supported by the CICYT project RTA2008-00027-00-00.

6. References

Aguera, A., Estrada, L.A.P., Ferrer, I., Thurman, E.M., Malato, S. & Fernandez-Alba, A.R. (2005). Application of time-of-flight mass spectrometry to the analysis of phototransformation products of diclofenac in water under natural sunlight, *Journal of Mass Spectrometry*, Vol. 40, pp. 908-915.

Ankumah, R.O., Dick, W.A. & McClung, G. (1995). Metabolism of carbamothioate herbicide, EPTC, by *Rhodococcus strain* JE1 isolated from soil, *Soil Science Society of American Journal*, Vol. 59, pp. 1071-1077.

Bachman, J. & Patterson, H.H. (1999). Photodecomposition of the carbamate pesticide carbofuran: kinetics and the influence of dissolved organic matter, *Environmental Science and Technology*, Vol. 33, pp. 874-881.

Barceló, D., Chiron, S., Fernandez, A., Valverde, A. & Alpendurada, M.F. (1996). Monitoring pesticides and metabolites in surface water and groundwater in Spain, *In* Meyer, M.T. & Thurman, E.M. (Eds.), *Herbicide metabolites in surface water and groundwater*, pp. 237-253, Oxford University Press, Washington D.C. (USA).

Belfroid, A.C., van Drunen, M., Beek, M.A., Schrap, S.M., van Gestel, C.A.M. & van Hattum, B. (1998). Relative risks of transformation products of pesticides for aquatic ecosystems, *Science of the Total Environment*, Vol. 222, pp. 167-183.

Berenzen, N., Lentzen-Godding, A., Probst, M., Schulz, H., Schulz, R. & Liess, M. (2005). A comparison of predicted and measured levels of runoff-related pesticide concentrations in small lowland streams on a landscape level, *Chemosphere*, Vol. 58, pp. 683-691.

Boule, P. (Ed.). (1999). *The handbook of environmental chemistry. Vol. II, Part L: Environmental photochemistry*, Springer-Verlag, Berlin (Germany).

Boxall, A.B.A., Sinclair, C.I., Fenner, K., Kolpin, D.W. & Maund, S.J. (2004). When synthetic chemicals degrade in the environment, *Environmental Science and Technology*, Vol. 1, pp. 369A-375A.

Bukovac, M.J., Sargent, J.A., Powell, R.G. & Blackman, G.E. (1971). Studies on foliar penetration. VII. Effects of chlorination on the movement of phenoxyacetic and benzoic acids through cuticles isolated from the fruits of *Lycopersimon esculentum L.*, *Journal of Experimental Botany*, Vol. 22, pp. 598-612.

Burgstahler, R.J. & Lichtenthaler, H.K. (1984). Inhibition by sethoxydim of phospho- and galactolipid accumulation in maize seedlings, *In* Siegenthaler, P.A. & Eichenbarger, W. (Eds.), *Structure, function and metabolism of plant lipids*, pp. 619-622, Elsevier Science, Amsterdam (The Netherlands).

Burton, J.D., Gronwald, J.W., Keith, R.A., Somers, D.A., Gengenbach, B.G. & Wyse, D.L. (1991). Kinetics of inhibition of acetyl-coenzyme A carboxylase by sethoxydim and haloxyfop, *Pesticide Biochemistry and Physiology*, Vol. 39, pp. 100-109.

Campbell, J.R. & Penner, D. (1985). Abiotic transformations of sethoxydim, *Weed Science*, Vol. 33, pp. 435-439.

Chow, P.N.P. & MacGregor, A.W. (1983). Effect of ammonium sulfate and surfactants on activity of the herbicide sethoxydim, *Journal of Pesticide Science*, Vol. 8, pp. 519-527.

Christoffers, M.J. & Kandikonda, A.V. (2006). Mechanisms of weed resistance to inhibitors of acetyl-CoA carboxylase, *Proceedings of Proceedings of 232nd ACS National Meeting*, San Francisco (USA), pp. AGRO-107.

Clark, J.R., Lewis, M.A. & Pait, A.S. (1993). Pesticide inputs and risks in coastal wetlands, *Environmental Toxicology and Chemistry*, Vol. 12, pp. 2225-2233.

Curtin, D.Y., Grubbs, E.J. & McCarty, C.G. (1966). Uncatalysed syn-anti isomerization of imines, oxime ethers, and haloimines, *Journal of American Chemical Society*, Vol. 88, pp. 2775-2786.

Dannenberg, A. & Pehkonen, S.O. (1998). Investigation of the heterogeneously calalyzed hydrolysis of organophosphorus pesticides, *Journal of Agricultural and Food Chemistry*, Vol. 46, pp. 325-334.

Delye, C., Straub, C., Chalopin, C., Matejicek, A., Michel, S. & Le Corre, V. (2003a). Résistance aux herbicides chez le vulpin. Un problème généralisé mais à gérer localement : sa gestion nationale semble peu envisageable, *Phytoma, la Défense des Végétaux*, Vol. 564, pp. 18-22.

Delye, C., Zhang, X.-Q., Chalopin, C., Michel, S. & Powles, S.B. (2003b). An isoleucine residue within the carboxyl-transferase domain of multidomain acetyl-coenzyme A carboxylase is a major determinant of sensitivity to aryloxyphenoxypropionate but not to cyclohexanedione inhibitors, *Plant Physiology*, Vol. 132, pp. 1716-1723.

Delye, C., Zhang, X.-Q., Michel, S., Matejicek, A. & Powles, S.B. (2005). Molecular bases for sensitivity to acetyl-coenzyme A carboxylase inhibitors in black-grass, *Plant Physiology*, Vol. 137, pp. 794-806.

Devine, M.D. & Shimabukuro, R.H. (1994). Resistance to acetyl-coenzyme A carboxylase inhibiting herbicides, *In* Powles, S.B. & Holtum, J.A. (Eds.), *Herbicide resistance in plants, biology and biochemistry*, pp. 141-169, Lewis Publishers, Boca Raton (USA).

Dimou, A.D., Sakkas, V.A. & Albanis, T.A. (2004). Photodegradation of trifluralin in natural waters and soils: degradation kinetics and influence of organic matter, *International Journal of Environmental Analytical Chemistry*, Vol. 84, pp. 173-182.

Dimou, A.D., Sakkas, V.A. & Albanis, T.A. (2005). Metolachlor photodegradation study in aqueous media under natural and simulated solar irradiation, *Jounal of Agricultural Food and Chemistry*, Vol. 53, pp. 694-701.

Elazzouzi, M., Bensaoud, A., Bouhaouss, A., Guittonneau, S., Dahchour, A., Meallier, P. & Piccolo, A. (1999). Photodegradation of imazapyr in the presence of humic substances, *Fresenius Environmental Bulletin*, Vol. 8, pp. 478-485.

Falb, L.N., Bridges, D.C. & Smith, A.E. (1990). Effects of pH and adjuvants on clethodim photodegradation, *Journal of Agricultural and Food Chemistry*, Vol. 38, pp. 875-878.

Falb, L.N., Bridges, D.C. & Smith, A.E. (1991). Separation of clethodim herbicide from acid and photodegradation products by liquid chromatography, *Journal of the Association of Official Analytical Chemists*, Vol. 74, pp. 999-1002.

Foy, C.L. (1993). Progress and developments in adjuvant use since 1989 in the USA, *Pesticide Science*, Vol. 38, pp. 65-76.

Foy, C.L. & Witt, H.L. (1992). Annual grass control in alfalfa (*Medicago sativa*) with postemergence graminicides, *Weed Technology*, Vol. 6, pp. 938-948.

García-Repetto, R., Martínez, D. & Repetto, M. (1994). The fluence of pH on the degradation kinetics of some organophosphorous pesticides in aqueous solutions, *Veterinary and Human Toxicology*, Vol. 36, pp. 202-204.

Green, P.G. & Young, T.M. (2006). Loading of the herbicide diuron into the California water system, *Environmental Engineering Science*, Vol. 23, pp. 545-551.

Harwood, J.L. (1999). Graminicides which inhibit lipid synthesis, *Pesticide Outlook*, Vol. 10, pp. 154-158.

Hashimoto, Y., Ishihara, K. & Soeda, Y. (1979a). Nature of the residue in soybean plant after treatment of alloxydim-sodium, *Journal of Pesticide Science*, Vol. 4, pp. 375-378.

Hashimoto, Y., Ishihara, K. & Soeda, Y. (1979b). Fate of alloxydim-sodium on or in soybean plants, *Journal of Pesticide Science*, Vol. 4, pp. 299-304.

Hatzios, K.K. (2001). Cases and mechanisms of resistance to ACCase-inhibiting herbicides, *In* Clark, J.M. & Yamaguchi, I. (Eds.), *Agrochemical Resistance: Extent, Mechanism, and Detection (ACS Symposium)*, pp. 135-149, Oxford University Press, Washington D.C. (USA).

Horowitz, M. (1980). Herbicidal treatments for control of Papaver somniferum L, *Bulletin on Narcotics*, Vol. 32, pp. 33-43.

Hsiao, A.I. & Smith, A.E. (1983). A root bioassay procedure for the determination of chlorsulfuron, diclofop acid and sethoxydim residues in soils, *Weed Research*, Vol. 23, pp. 231-236.

Hu, J.-Y., Aizawa, T. & Magara, Y. (1999). Analysis of pesticides in water with liquid chromatography/atmospheric pressure chemical ionization mass spectrometry, *Water Research*, Vol. 33, pp. 417-425.

Ibáñez, M., Sancho, J.V., Pozo, O.J. & Hernández, F. (2004). Use of quadrupole time-of-flight mass spectrometry in environmental analysis: elucidation of transformation products of triazine herbicides in water after UV exposure, *Analytical Chemistry*, Vol. 76, pp. 1328-1335.

Iwataki, I. (1992). Cyclohexanedione herbicides: their activities and properties, *In* Draber, W. & Fujita, T. (Eds.), *Rational approaches to structure, activity and ecotoxicology of agrochemicals*, pp. 397-426, CRC Press, Boca Raton (USA).

Iwataki, I. & Hirono, Y. (1978). The chemical structure and herbicidal activity of alloxydim-sodium and related compounds, *In* Geissbühler, H., Brooks, G.T. & Kearney, P.C. (Eds.), *Advances in Pesticide Science, Fourth International Congress on Pesticide Chemicals*, pp. 235-243, Pergamon Press, Zurich (Switzerland).

Junghans, M., Backhaus, T., Faust, M., Scholze, M. & Grimme, L.H. (2006). Application and validation of approaches for the predictive hazard assessment of realistic pesticide mixtures, *Aquatic Toxicology*, Vol. 76, pp. 93-110.

Kolpin, D.W., Schnoebelen, D.J. & Thurman, E.M. (2004). Degradates provide insight to spatial and temporal trends of herbicides in ground water, *Ground Water*, Vol. 42, pp. 601-608.

Koskinen, W.C., Reynolds, K.M., Buhler, D.D., Wyse, D.L., Barber, B.L. & Jarvis, L.J. (1993). Persistence and movement of sethoxydim residues in three Minnesota soils, *Weed Science*, Vol. 41, pp. 634-640.

Krogh, K.A., Halling-Sorensen, B., Mogensen, B.B. & Vejrup, K.V. (2003). Environmental properties and effects of nonionic surfactant adjuvants in pesticides: a review, *Chemosphere*, Vol. 50, pp. 871-901.

Kuk, Y.-I., Burgos, N.R. & Talbert, R.E. (2000). Cross- and multiple resistance of diclofop-resistant *Lolium spp.*, *Weed Science*, Vol. 48, pp. 412-419.

Leach, G.E., Devine, M.D., Kirkwood, R.C. & Marshall, G. (1995). Target enzyme-based resistance to acetyl-coenzyme A carboxylase inhibitors in *Eleusine Indica*, *Pesticide Biochemistry and Physiology*, Vol. 51, pp. 129-136.

Li, H.-Y. & Foy, C.L. (1999). A biochemical study of BAS 517 using excised corn and soybean root systems, *Weed Science*, Vol. 47, pp. 28-36.

Lykins, B.W., Koffskey, W.E. & Miller, R.G. (1986). Chemical products and toxicologic effects of disinfection, *Journal of American Water Works Association*, Vol. 78, pp. 66-75.

Magara, Y., Aizawa, T., Matumoto, N. & Souna, F. (1994). Degradation of pesticides by chlorination during water purification, *Water Science and Technology*, Vol. 30, pp. 119-128.

Marcheterre, L., Choudhry, G.G. & Webster, G.R.B. (1988). Environmental photochemistry of herbicides, *Reviews of Environmental Contamination and Toxicology*, Vol. 103, pp. 61-126.

Marles, M.A.S., Devine, M.D. & Hall, J.C. (1993). Herbicide resistance in *Setaria viridis* conferred by a less sensitive form of acetyl-coenzyme A carboxylase, *Pesticide Biochemistry and Physiology*, Vol. 46, pp. 7-14.

McInnes, D., Harker, K.N., Blackshaw, R.E. & Born, W.H.V. (1992). The influence of ultraviolet light on the phytotoxicity of sethoxydim tank mixtures with various adjuvants, *In* Foy, C.L. (Ed.), *Adjuvants for agrichemicals*, pp. 205-213, CRC Press, Boca Raton (USA).

McMullan, P.M. (1996). Grass herbicide efficacy as influenced by adjuvant, spray solution pH, and ultraviolet light, *Weed Technology*, Vol. 10, pp. 72-77.

Nalewaja, J.D., Matysiak, R. & Szelezniak, E. (1994). Sethoxydim response to spray carrier chemical properties and environment, *Weed Technology*, Vol. 8, pp. 591-597.

Neilson, A.H. & Allard, A.-S. (2008). *Environmental degradation and transformation of organic chemicals*, CRC Press, Boca Raton (USA).

Ono, S., Shiotani, H., Ishihara, K., Tokieda, M. & Soeda, Y. (1984). Degradation of the herbicide alloxydim-sodium in soil, *Journal of Pesticide Science*, Vol. 9, pp. 471-480.

Park, H. & Choi, W. (2003). Visible light and Fe(III) mediated degradation of acid orange 7 in the absence of H_2O_2, *Journal of Photochemistry and Photobiology A*, Vol. 159, pp. 241-247.

Park, K.W. & Mallory-Smith, C.A. (2004). Physiological and molecular basis for ALS inhibitor resistance in *Bromus tectorum* biotypes, *Weed Research*, Vol. 44, pp. 71-77.

Peñuela, G.A., Ferrer, I. & Barceló, D. (2000). Identification of new photodegradation by-products of the antifouling agent irgarol in seawater samples, *International Journal of Environmental Analytical Chemistry*, Vol. 78, pp. 25-40.

Price, L.J., Herbert, D., Cole, D.J. & Harwood, J.L. (2003). Use of plant cell cultures to study graminicide effects on lipid metabolism, *Phytochemistry*, Vol. 63, pp. 533-541.

Reckhow, D.A. & Singer, P.C. (1990). Chlorination by-products from drinking waters: from formation potential to finished water concentrations, *Journal of the American Water Works Association*, Vol. 82, pp. 173-180.

Reineke, N., Bester, K., Huhnerfuss, H., Jastorff, B. & Weigel, S. (2002). Bioassay-directed chemical analysis of river Elbe surface water including large volume extractions and high performance fractionation, *Chemosphere*, Vol. 47, pp. 717-723.

Rendina, A.R., Craig-Kennard, A.C., Beaudoin, J.D. & Breen, M.K. (1990). Inhibition of acetyl-coenzyme A carboxylase by two classes of grass-selective herbicides, *Journal of Agricultural and Food Chemistry*, Vol. 38, pp. 1282-1287.

Richardson, S.D. (2006). Environmental mass spectrometry: emerging contaminants and current issues, *Analytical Chemistry*, Vol. 78, pp. 4021-4045.

Richardson, S.D. (2007). Water analysis: emerging contaminants and current issues, *Analytical Chemistry*, Vol. 79, pp. 4295-4324.

Richardson, S.D. (2009). Water analysis: emerging contaminants and current issues, *Analytical Chemistry*, Vol. 81, pp. 4645-4677.

Ritz, C., Cedergreen, N., Jensen, J.E. & Streibig, J.C. (2006). Relative potency in nonsimilar dose-response curves, *Weed Science*, Vol. 54, pp. 407-412.

Roberts, T.R. (Ed.). (1998). *Metabolic pathways of agrochemicals*, Royal Society of Chemistry, Cambridge (UK).

Rodríguez-Mozaz, S., López de Alda, M.J. & D., B. (2007). Advantages and limitations of on-line solid phase extraction coupled to liquid chromatography-mass spectrometry technologies versus biosensors for monitoring of emerging contaminants in water, *Journal of Chromatography A*, Vol. 1152, pp. 97-115.

Sakkas, V.A., Lambropoulou, D.A. & Albanis, T.A. (2002a). Study of chlorothalonil photodegradation in natural waters and in the presence of humic substances, *Chemosphere*, Vol. 48, pp. 939-945.

Sakkas, V.A., Lambropoulou, D.A. & Albanis, T.A. (2002b). Photochemical degradation study of irgarol 1051 in natural waters: influence of humic and fulvic substances on the reaction, *Journal of Photochemistry and Photobiology A*, Vol. 147, pp. 135-141.

Sandín-España, P. (2004). *Thesis: Estudio de la degradación y análisis de herbicidas ciclohexanodionas en agua clorada*, Universidad Autónoma de Madrid, Madrid (Spain).

Sandín-España, P., González-Blázquez, J.J., Magrans, J.O. & García-Baudín, J.M. (2002). Determination of herbicide tepraloxydim and main metabolites in drinking water by solid-phase extraction and liquid chromatography with UV detection, *Chromatographia*, Vol. 55, pp. 681-686.

Sandín-España, P., Llanos, S., Magrans, J.O., Alonso-Prados, J.L. & García-Baudín, J.M. (2003). Optimization of hydroponic bioassay for herbicide tepraloxydim by using water free from chlorine, *Weed Research*, Vol. 43, pp. 451-457.

Sandín-España, P., Magrans, J.O. & García-Baudín, J.M. (2005a). Study of clethodim degradation and by-product formation in chlorinated water by HPLC, *Chromatographia*, Vol. 62, pp. 133-137.

Sandín-España, P., Santín, I., Magrans, J.O., Alonso-Prados, J.L. & García-Baudín, J.M. (2005b). Degradation of alloxydim in chlorinated water, *Agronomy for Sustainable Development*, Vol. 25, pp. 331-334.

Santín-Montanya, I., Sandín-España, P., García Baudín, J.M. & Coll-Morales, J. (2007). Optimal growth of *Dunaliella primolecta* in axenic conditions to assay herbicides, *Chemosphere*, Vol. 66, pp. 1315-1322.

Santoro, A., Scopa, A., Bufo, S.A., Mansour, M. & Mountacer, H. (2000). Photodegradation of the triazole fungicide hexaconazole, *Bulletin of Environmental Contamination and Toxicology*, Vol. 64, pp. 475-480.

Santos, T.C.R., Rocha, J.C., Alonso, R.M., Martínez, E., Ibáñez, C. & Barceló, D. (1998). Rapid degradation of propanil in rice crop fields, *Environmental Science and Technology*, Vol. 32, pp. 3479-3484.

Sanz-Asencio, J., Plaza-Medina, M. & Martínez-Soria, M.T. (1997). Kinetic study of the degradation of ethiofencarb in aqueous solutions, *Pesticide Science*, Vol. 50, pp. 187-194.

Schwarzenbach Rene, P., Gschwend, P.M. & Imboden, M.D. (2002). *Environmental organic chemistry* (2nd), John Wiley & Sons, New Jersey (USA).

Secor, J. & Cseke, C. (1988). Inhibition of acetyl-CoA carboxylase activity by haloxyfop and tralkoxydim, *Plant Physiology*, Vol. 86, pp. 10-12.

Seefeldt, S.S., Gealy, D.R., Brewster, B.D. & Fuerst, E.P. (1994). Cross-resistance of several diclofop-resistant wild oat (*Avena fatua*) biotypes from the Willamette Valley of Oregon, *Weed Science*, Vol. 42, pp. 430-437.

Sevilla-Morán, B. (2010). *Thesis: Estudio de la fotodegradación de herbicidas ciclohexanodionas en medio acuoso*, Universidad Autónoma de Madrid, Madrid (Spain).

Sevilla-Morán, B., Alonso-Prados, J.L., García-Baudín, J.M. & Sandín-España, P. (2010a). Indirect photodegradation of clethodim in aqueous media. By-product identification by quadrupole time-of-flight mass spectrometry, *Journal of Agricultural and Food Chemistry*, Vol. 58, pp. 3068-3076.

Sevilla-Morán, B., Mateo-Miranda, M., López-Goti, C., Alonso-Prados, J.L. & Sandín-España, P. (2011). Photodegradation of profoxydim in natural waters. Comparative study of the photolytic behaviour of the active substance and its formulation Aura®, *Proceedings of Proceedings of XIV Symposium in Pesticide Chemistry. Pesticides in the environment: fate, modelling and risk mitigation*, Piacenza (Italy).

Sevilla-Morán, B., Mateo-Miranda, M.M., Alonso-Prados, J.L., García-Baudín, J.M. & Sandín-España, P. (2010b). Sunlight transformation of sethoxydim-lithium in natural waters and effect of humic acids, *International Journal of Environmental Analytical Chemistry*, Vol. 90, pp. 487-496.

Sevilla-Morán, B., Sandín-España, P., Vicente-Arana, M.J., Alonso-Prados, J.L. & García-Baudín, J.M. (2008). Study of alloxydim photodegradation in the presence of natural substances: Elucidation of transformation products, *Journal of Photochemistry and Photobiology A*, Vol. 198, pp. 162-168.

Shoaf, A.R. & Carlson, W.C. (1986). Analytical techniques to measure sethoxydim and breakdown products, *Weed Science*, Vol. 34, pp. 745-751.

Shoaf, A.R. & Carlson, W.C. (1992). Stability of Sethoxydim and its degradation products in solution, in soil, and on surfaces, *Weed Science*, Vol. 40, pp. 384-389.

Shukla, A., Nycholat, C., Subramanian Mani, V., Anderson Richard, J. & Devine Malcolm, D. (2004). Use of resistant ACCase mutants to screen for novel inhibitors against resistant and susceptible forms of ACCase from grass weeds, *Journal of Agricultural and Food Chemistry*, Vol. 52, pp. 5144-5150.

Smith, A.E. & Hsiao, A.I. (1983). Persistence studies with the herbicide sethoxydim in prairie soils, *Weed Research*, Vol. 23, pp. 253-257.

Soeda, Y., Ishihara, K., Iwataki, I. & Kamimura, H. (1979). Fate of a herbicide ^{14}C-alloxydim-sodium in sugar beets, *Journal of Pesticide Science*, Vol. 4, pp. 121-128.

Somasundaram, L. & Coats, J.R. (Eds.). (1991). *Pesticide transformation products. Fate and significance in the environment*, Oxford University Press, Washington D.C. (USA).

Srivastava, A. & Gupta, K.C. (1994). Dissipation of tralkoxydim in water-soil system, *Journal of Pesticide Science*, Vol. 19, pp. 145-149.

Stoltenberg, D.E. & Wiederholt, R.J. (1995). Giant foxtail (*Setaria faberi*) resistance to aryloxyphenoxypropionate and cyclohexanedione herbicides, *Weed Science*, Vol. 43, pp. 527-535.

Tal, A. & Rubin, B. (2004). Molecular characterization and inheritance of resistance to ACCase-inhibiting herbicides in *Lolium rigidum*, *Pest Management Science*, Vol. 60, pp. 1013-1018.

Tchaikovskaya, O., Sokolova, I., Svetlichnyi, V., Karetnikova, E., Fedorova, E. & Kudryasheva, N. (2007). Fluorescence and bioluminescence analysis of sequential UV-biological degradation of p-cresol in water, *Luminescence*, Vol. 22, pp. 29-34.

Tixier, C., Meunier, L., Bonnemoy, F. & Boule, P. (2000). Phototransformation of three herbicides: chlorbufam, isoproturon, and chlorotoluron. Influence of irradiation on toxicity, *International Journal of Photoenergy*, Vol. 2, pp. 1-8.

Tomlin, C.D.S. (Ed.). (2006). *The pesticide manual: a world compendium*, BCPC Publications, Hampshire (UK).

Tuxhorn, G.L., Roeth, F.W., Martin, A.R. & Wilson, R.G. (1986). Butylate persistence and activity in soils previously treated with thiocarbamates, *Weed Science*, Vol. 34, pp. 961-965.

Vialaton, D. & Richard, C. (2002). Phototransformation of aromatic pollutants in solar light: photolysis versus photosensitized reactions under natural water conditions, *Aquatic Sciences*, Vol. 64, pp. 207-215.

Walker, K.A., Ridley, S.M., Lewis, T. & Harwood, J.L. (1988). Fluazifop, a grass-selective herbicide which inhibits acetyl-CoA carboxylase in sensitive plant species, *Biochemical Journal*, Vol. 254, pp. 307-310.

Walter, H. (2001). Profoxydim: development of a herbicide from laboratory to field, *In* Prado, R.D. & Jorrín, J.V. (Eds.), *Uso de herbicidas en la agricultura del siglo XXI*, pp. 19-30, Servicio Publicaciones Universidad de Córdoba, Córdoba (Spain).

Wiederholt, R. & Stoltenberg, D.E. (1995). Cross-resistance of a large crabgrass (*Digitaria sanguinalis*) accession to aryloxyphenoxypropionate and cyclohexanedione herbicides, *Weed Technology*, Vol. 9, pp. 518-524.

7

Persistence of Herbicide Sulfentrazone in Soil Cultivated with Sugarcane and Soy and Effect on Crop Rotation

Flávio Martins Garcia Blanco[1], Edivaldo Domingues Velini[2]
and Antonio Batista Filho[1]
[1]*Instituto Biológico de São Paulo, Centro Experimental, Campinas*
[2]*Universidade Estadual Paulista - UNESP, Faculdade de Ciências Agronômicas, Botucatu*
Brazil

1. Introduction

In the deployment of an agricultural area through a cultivation system, there are serious and significant changes in geomorphic subsystems, edaphic and biological, making them simpler (agroecosystem), compared with the ecosystem, this, a more complex system. This change results in drastic reduction of the self-regulatory system, making it more unstable and susceptible to energy inputs. One major consequence of this transformation is the excessive increase of some species of insects, microorganisms and nematodes populations, and wild plants in such a way as to significantly impair the production, making uneconomical the productive units, so, they are named as agricultural pests.

When the wild plants interfere with cultivated plants, specifically, they become weeds, which unlike others agricultural pests, they are by nature always present in agroecosystems, they are difficult to control and are directly (competition, allelopathy, etc.) or indirectly (reservoir of pathogens, insects and nematodes) responsible for the drastic decrease in economic production of crops.

For a long time, agricultural researches have demonstrated that the control of weeds in various agroecosystems is a key factor for the success of crop production. All the technological development in crop science, nutrition, or breeding, can be compromised if the weeds were not controlled. The weed control is done by combining several methods, such as preventive, cultural, and chemical weeding, this, by the use of herbicides.

Herbicides are chemicals used to eliminate plants. They are applied in suitable doses directly on the vegetation for foliar absorption (post-emergence treatment), or on the soil for absorption by the plant tissues formed after the seed germination, before the plant emergence from the soil surface (pre-emergence treatment). They are generally used to control of weeds in different agro-ecosystems, or in any other favorable ecological niches of these organisms: wasteland, margins of highways, railroad beds, parking lots, and aquatic environments.

To selection of which herbicide will by used to weed control, you should always have an ecological focus using this agronomic technique aiming the maximum production. This duality, choice, besides the type, dose, number and mode of application, should always seek the dichotomy of maximum efficiency and minimum environmental impact, thus maximizing the benefits of their use and minimizing their environmental and toxicological risks

Even so, the use of these chemicals is not without risks, with possible presence of residues in agroecosystems that can cause toxicity to susceptible plants used in rotation with the crops originally treated with the herbicides.

Brazil is a country where agricultural production is has world importance, notably in the growing of sugarcane and soybean, which since the end of the last century, with the implementation of the program using ethanol to replace gasoline, and the cultivation of soybeans in the Brazilian Cerrado (savanna), their productions and productivities have increased significantly each year.

These two crops are cultivated in extensive areas, because of their ease of use and performance; and the use of herbicides in these crops is intensive and in many cases, especially in sugarcane, in that most herbicides have long residual power, persisting in soil for a very long time. Thus, these two crops, sugar cane and soybeans, must be those with the greatest potential risk of occurrence of problems related to the persistence of herbicide molecules in the soil for a longer period than it is desirable, with the possibility of causing environmental contamination and phytotoxicity to sensitive crops used in rotation, especially with soybean that has a shorter cycle.

These themes are linked with the herbicide ecotoxicology, and only in 1960 began the interest in studies on the ecological effects of chemicals, when then the society begins to worry about their effects on environmental contamination due to, primarily, the world press reports on the effects of insecticides for agricultural use on the wildlife. A classic example occurred in the 60's in Mississippi and Atchafalaya rivers, United States, resulting in the deaths of ten millions fishes from water polluted with the insecticide endrin (Madhun & Freed, 1990). In 1962, there was a great repercussion around the world, with the release of the book "Silent Spring", written by Rachel Carson (1964), which projected an obscure future for planet Earth, if the man did not stop using pesticides in an indiscriminate way.

Soon after this season, in 1975, it was started the development of Ecotoxicology, a branch of science created by Rene Thruhaut in Paris (Astolfi et al, 1984), studying the mechanisms of environmental contamination by natural and artificial chemicals (xenobiotics) as well as the action   such substances and their effects on living beings that inhabit the biosphere. Ecotoxicology is a natural extension of Toxicology.

One test that has contributed to understanding the behavior of pesticides are the ecotoxicological field tests to verify and monitor the persistence, accumulation, degradation and leaching of these products in soil.

2. Behaviour of herbicides in soils, especially sulfentrazone

The agricultural soil is the final destination of a large number of herbicides, either when they are applied directly to the soil or on the shoots of plants (Walker, 1987). When the herbicides, reach the ground, interacting with the environment, their fate is governed by three general types of processes: physical (sorption-desorption, volatilization, leaching by water erosion and transportation along the ground by wind and water); chemicals (photodecomposition, sorption, chemical reactions with the soil constituents) and biological (represented by the microbial decomposition of the molecule and removal of soil by plants), Sheets, 1970, Blanco, 1979.

All these processes are described in the opinion of Briggs (1969, 1976, 1981), Blanco (1979), Walker (1983), Walker & Allen (1984) and Velini (1992), dependent on the chemical and physical soil nature and climatic conditions, particularly temperature and soil moisture, and soil characteristics (texture, structure, content and nature of colloid, pH, temperature, humidity, and others). The chemical nature of each herbicide, in turn, is a function of its molecular structure, molecule ionization, water solubility, lipid solubility, polarity and volatilization of the molecule. On the other hand, several external factors may play an important role in herbicide-soil interactions, such as dose and application mode, the herbicide formulation and soil microbial community

It is understood by adsorption to the accession of a molecule, ion or particle of the surface in any other particle, resulting from the interaction of a force field emanating from the adsorbent surface (clay and organic matter) and the surface of the adsorbate (in this case a herbicide). Herbicide Particles can also be absorbed by soil colloids. In 1994, Harper pondered the difficulty of distinguishing between the phenomena of absorption and adsorption, suggesting the use of the term sorption, which refers both cases.

Briggs (1969, 1976) reported that the extent and intensity of the processes involved in the phenomenon of sorption / desorption depend largely on molecular properties (physical and chemical) of the herbicide and the temperature, humidity, and soil pH and colloids. Herbicides can be molecular, weak acids or bases with their ionization depending on the soil pH, when the herbicide is the value of its ionization constant (pK), near the soil pH, the predominant form (molecular or ionic) , can vary greatly with a slight change in soil pH, Figure 1 illustrates it for sulfentrazone (pk = 6.56), characterized with weak acid.

Figure 1 shows the effect of soil pH rate on the percentage of the herbicide shape, ionized (anion) or molecular, and helps to understand the article of Grey et al. (1997), who analyzed the sorption of sulfentrazone when applied in multiple doses, varying the soil pH index, note that this characteristic has very significantly influence on the phenomenon of herbicide sorption to soil colloids, that decreases in response to a increasing pH, especially when this increase occurs above the pK of the herbicide (6.56), that because of the increased concentration of ionized form (anion) and a decrease in molecular form, in reverse, there is increased sorption to colloids when the index decreases, especially when this value below the pK of sulfentrazone.

It should be noted that this proportionality has a logarithmic behavior, so that a small variation in pH values leads to a major shift in the predominant form of the herbicide (molecular or ion).

This is a reality for Brazilian soils that have a pH range that include the pK of sulfentrazone (6.56), theoretically in two soils in the same climatic region with soil texture and organic matter equal but with the soil pH ranging from 5.5 to 7.2, the percentage of ionization would vary from 8.01 and 81.32% respectively, and may thus influence significantly the sorption of herbicides to soil colloids and in consequence, there are different persistence just for this variation in soil pH.

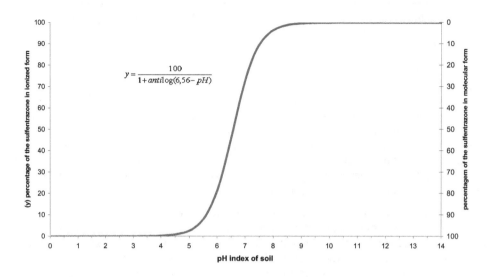

Fig. 1. Logistic model showing the ionization variation and molecular percentage of sulfentrazone (pk = 6.56), depending the content of the soil pH.

It should be noted that at low pH, even with a higher proportion of ionized form in the proton hydrogen (H^+) in soil solution competes with the ionized form of the herbicide by the site of sorption to soil colloids, so in many cases there is a greater sorption when soil pH is equal to the pK of the acid herbicide.

Reddy & Locke (1998), studying the sulfentrazone interaction on soil microorganisms using the labeled carbon method in the radical of the phenol molecule, found that after 77 incubation days, only 2.1% of 14 C-sulfentrazone were degraded by respiration edaphic to 14 CO_2, concluding that the native population of microorganisms edaphic had a small adaptation to the herbicide, could not dispel and degrade it, suggesting that this pathway is not the most efficient for their removal of the environment.

Chemical dissipation of the herbicide sulfentrazone is not yet fully understood, we know that does not show hydrolysis, photolysis are stable when applied to the soil, but extremely sensitive to this process in the water (EPA, 2011, Reddy & Locke 1998). All of these processes and factors influencing the sulfentrazone soil persistence.

3. Herbicides persistence in soils

Seeds of agricultural crops, when thrown to the ground, germinate at once and have slow initial growth in relation to the weeds. Herbicides applied directly on the ground need to persist in action on the ground with multiple streams of weed emergence (residual power) until the final limit of the critical competition period that for some crops is relatively long in Brazil.

For example, for the soybean crop in Brazilian conditions, this period is 30 to 50 days after emergence, while the sugarcane is planted in southeastern Brazil, at different periods during the year, the critical competition periods of varies according to these periods, a shorter period for spring and summer plantings, 15 to 60 days, and a longer, 60 to 90 days after emergence of seedlings, for autumn / winter planting because of drought that halts the growth of the plants until the next rain season (Blanco et. al 1978, 1979, 1981), so the herbicides used in this condition will persist for a long time with residual action to weeds in order to control their first flows allowing the plant to grow and expanding their leaves from parallel planting lines, shading the spacing of planting, controlling these plants through of light competition.

By definition, the persistence of herbicides in soil can be of three types, according to the method and objective of its biological, agronomic and chemical persistence determination. The biological persistence is determined by biological methods (bioassays) the time of the residue effect on living beings (bioactivity), in which case the agronomic persistence is determined using plant test, a biological persistence is individualized, therefore, measurements of the time that the herbicide residue with activity remains in soil can affect plants grown in a system of succession or rotation of crops. The chemical persistence concerns at time of the residue remains in soil that can be detected by chemical or radiometric methods.

The chemical methods (residues analysis) quantify the level of the herbicide while the biological (bioassay) qualify the presence of this, and in many cases this method is more sensitive, as all methodologies, advantages and limitations, and their use depend on the purpose of each test. Figure 2, show inferences about the persistence of herbicides in soil and their action.

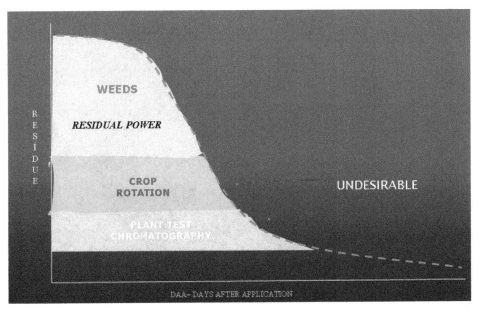

Fig. 2. Show a theoretical model by a logistic function, emphasis areas for action or method capable of detecting the herbicide, thus illustrating the decreasing variation of the residual herbicide in soil in time function (persistence).

We observe that the highest concentrations of the herbicide residues in soil correspond to the desirable effect of weed control (residual power). Over time, the herbicide concentration in soil decreases to levels that no weed control, but its concentration can affect crops in succession to that in which the herbicide was applied originally; their presence can be detected by symptoms of phytotoxicity expressed in these crops when they are sensitive. From this time, the residual level of the herbicide is very low, being detected only by plants with extremely susceptibility to the herbicide (plant test) or by analytical methods (eg chromatography). From this point, the concentration of herbicide in the soil is so low that current technology can not detect its presence on the ground.

Thus, the ideal herbicide would be that for which there is a coincidence of the final period of their persistence with the final period of the critical competition period the crop, a situation that unfortunately does not occur for current herbicides.

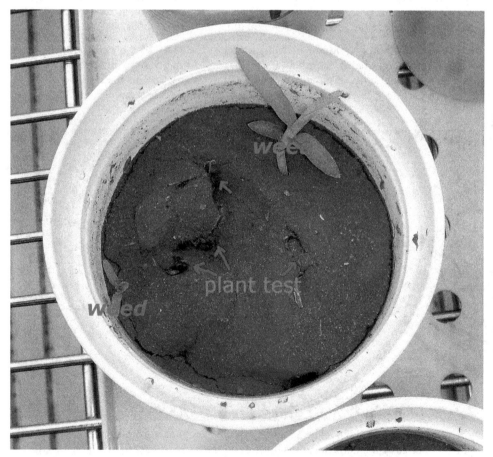

Fig. 3. Bioassay showing different action of the herbicide. Photo: Flávio M. G. Blanco.

Figure 3 explains the difference between persistence and residual power of the herbicide. Although the herbicide is present (persistent) in the soil, because it affects the plant test, does not affect the weeds, in this case the herbicide persists in the soil, but without power or residual effect on weeds.

Aiming aspects of selectivity for crops in succession to determine the persistence using the methodology to plant test is more efficient because the residual level in the soil did not affect the plant test, the more sensitive and also not cause injury to succession crops.

It should be emphasized that the use of results from ecotoxicological research, with aim to study the persistence of herbicides in soils in other countries (not Brazil) should be used with reservation, since different soil and climatic conditions often change the action of herbicides in soil, moreover, even when considering the Brazilian territory, it is difficult to extrapolate, for example, the results obtained in the Southeast to the Northeast Region.

In Brazil, the research works of herbicides in soils in agroecosystems are still few, especially when obtained under natural conditions of cultivation, such as the determination of persistence in the soil sulfentrazone, as well as its effects on cultures in a system of succession

3.1 Soil persistence of sulfentrazone in Brazilian conditions

3.1.1 Persistence determination by bioassays

Sulfentrazone herbicide is registered in Brazil for soybean, sugarcane, coffee and citrus. Belongs to the aryl-triazolinonas herbicide group, solubility of 780 mg L $^{-1}$ (pH $_7$), vapor pressure 1.10^{-9} mmHg (25 C°), dissociation constant (pK) 6.56 and partition coefficient (Kow $_{(pH7)}$ 9.8), belongs to group aryl triazolinonas chemical, mode of action is to the destruction of cell membranes by inhibiting the enzyme Protox is the accumulation of protoporphyrin IX causing peroxidation of O_2 and consequently the destruction of cell membranes.

Rossi et al. (2003) using bioassays, evaluated the leaching of sulfentrazone (0.86 kg ha^{-1}), in two Brazilian soils (Red Nitosol and Neosoil Quartzarênico), in PVC columns; 0.10 and 0.50 m (diameter and length) using sorghum (*Sorghum bicolor*) under different rainfall regimes for 15 days, determined that over rainfall of 90 mm, the herbicide was detected by 7.5 and 12.5 respectively in Red Nitosol and Neosoil Quartzarênico soil.

These data are consistent with Vivian et al. (2006), also in brazilian conditions, investigating the actions of sulfentrazone, up to 20 cm deep, applied at the dosage of 0.9 kg ha^{-1} in sugarcane crop, determined with bioassays using sorghum as plant test, the leaching of herbicide was significant only on 0-10 cm layer of soil and persisted up to 467 days. However, when reapplying the herbicide in sugarcane crop, the order of persistence could not be determined, because until the last evaluation (640 days) the herbicide still persisted in the soil. In the same work, it was given the relationship of sorption (RA) in 3.6, indicating that this herbicide, when sorbed onto colloids, has a tendency for slow release into soil solution.

For new herbicides the difficulty is to determine which plant test should be used, especially when the group is composed of a few chemical elements. To determine the sensitivity, several mathematical models can be tested, to quantify and explain the phenomenon:

exponential, reciprocal, logistic, quadratic, cubic, Gompertz, exponential, quadratic and cubic, and others. The important aspect is not to analyze only the mathematical aspect of the model, such as the coefficient of determination that reflects the goodness of fit, but also the logic of biological phenomena, thus the choice of model will have a higher probability of success.

For Streibig et al. (1988, 1995), when compared with other methods for the detection of residues in agroecosystems, such as chromatography, the mathematical modeling studies using bioassay methods, is not fully standardized yet. There are several types of models, in the literature, to estimate the phenomenon, however, the authors indicate the logistic or log-logistic model as the most appropriate to explain the dependence of plant development with herbicide dose variation, and especially for calculating the RG 50 (herbicide dose that reduces growth by 50% of the plant). The same opinion is shared by Seefeldt et al. (1995).

Thus, Blanco (2002), using the bioassay method, tested two plant species, candidates as test plant for sulfentrazone herbicide, evaluating the biological responses of sorghum (*Sorghum bicolor*), cv AG 2002 and sugar beet (*Beta vulgaris*), cv. Early Wonder, exposed to a series of herbicide dilutions, growing for 14 days in 300 ml plastic cups, without percolation, with 250 g of soil with medium texture, kept at 80% of field capacity by daily watering in a fitotron Conviron® model PVG386, at 20° C, 70-80% relative humidity and photoperiod of 16 hours, light intensity of 35,400 lumens m-2, to determine the dose that reduces 50% of their fresh weight of the epigeal part (RG 50%), using logistic function as the mathematical model (Fig. 4).

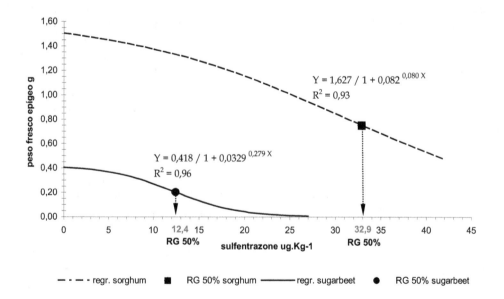

Fig. 4. Effect of increasing doses of sulfentrazone on epigeal fresh weight of sugar beet and sorghum and determination of the RG (50%) of each plant by logistic regression models, mean values of 10 replicates. (Blanco, 2002).

The analysis of the sensitivity of the plant test, using the methodology to evaluating the "fresh epigeal mass" instead of "epigeal dry mass," is due to feature of the herbicide, which belongs to the family of aryl triazolynonas, and is a desiccant, so, the methodology of evaluating the "epigeal dry mass," would tend to remove all differences between treatments, making all of them similar to the control, considering that a turgid leaf, when dried, can present the same symptoms (chlorosis, necrosis and desiccation) as a leaf treated with herbicide. Drying the plant material, it decreases the variability between treatments and, consequently, increasing the difficulty in finding significant differences between them, either by treatments analysis of variance (F test) or by an independent medium test

As observed in Figure 4, the logistic model adapts well to explaining the phenomenon, either by their biological logic - gradual reduction in mass with the increase of doses tested, as well as mathematical logic - high coefficients of determination, 0.93 and 0.96, showing excellent fit of the data to the chosen model. It should be noticed that to obtain the mathematical models, we used an appropriate number of observations, depending upon the extent of the analyzed data, fifteen and ten grain sorghum and sugar beet, respectively with a spacing between them of 3 µg of sulfentrazone, contributing to good accuracy of the obtained models

It is emphasized that the high light intensity, 35,400 lumens m^{-2}, and the photoperiod of 16 hours, for the bioassay, increased the sensitivity of the test plant, because sulfentrazone has its mode of action activated by the light intensity.

In figure 4, it is shown that both plants were very sensitive to the herbicide, notably beets with the highest sensitivity with RG (50%) equal to 12.4 µg.kg^{-1}, less than the half of the value observed to sorghum. Thus, the beet was used as the test plant for biological assays to determine the persistence of sulfentrazone in sugarcane and soybean crops in the proceedings of Blanco & Velini, 2005 and Blanco et al. 2010, described below.

3.1.2 Soil persistence of sulfentrazone applied in sugarcane crops.

Blanco et al. 2010, studying the sulfentrazone (0.6 and 1.2 kg ha $^{-1}$) in Brazilian conditions, soil pH 6.4 and organic matter, 11 g dm-3, using the bioassay method evaluating herbicide persistence under field conditions, the sugarcane crop until 704 days after treatment application, obtained the following biological responses for the plant test, described in the Figure 5.

During the test period, the soil was sampled in 23 seasons, immediately after application (0), up to 704 DAT, where the soil was sampled to determine the persistence by bioassay method, using sugar beet as test plant, growing for 14 days in a phytotron set at 20 ° C, 70-80% relative humidity and photoperiod of 16 hours, with light intensity of 35,400 lumens $m^{-2.}$

The climatic condition during the test is represented by figure 6.

The doses evaluated had different behavior dose of 0.6 kg ha^{-1} increased gradually from fresh epigeal not deferred test $t_{(p <5\%)}$ from the control 601 DAT by the end of sampling at 704 DAT. The dose 1.2 kg ha^{-1}, the averages of fresh epigeal also increased and differed significantly from the control, until the end of sampling.

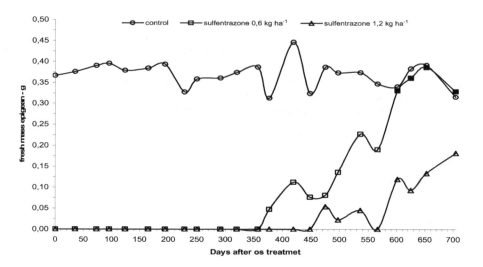

Fig. 5. Temporal variation of fresh epigeal a function of days after treatment, ■ represents no significant difference (p> 0.005), compared with the control. Each symbol represents the mean value of 5 replicates (Blanco et al. 2010).

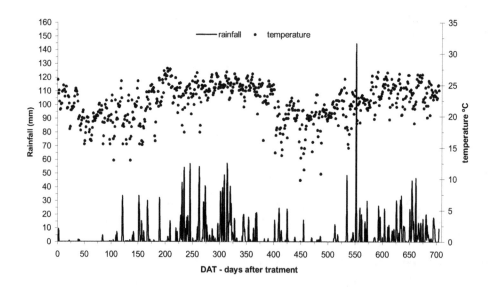

Fig. 6. Daily rainfall and average air temperature in the period from 30/03/2000 to 04/03/2002. (Blanco et al. 2010).

Through these values, we can determine the final limit of the persistence of the herbicide at 0.6 kg ha^{-1}, 601 DAT, the highest dose was found this persistence has lasted over 103 days, signaling that the sulfentrazone persistence when applied at 1.2 kg ha^{-1}, may exceed 704 DAT.

The timing of herbicide in early fall was characterized by a gradual decrease in rainfall and water content in soil, which benefited the herbicide sorption to soil colloids and discouraging the development of the microbial community, participatory processes dissipation herbicides (Alexander 1965; Weber 1970), it favored the retention of the herbicide in the soil, which would explain the lack of plant development in the first test ratings, which occurred only after 150 DAT, with the onset of rains and the increase in temperature, remaining until 450 DAT.

This condition favored the development of the microbial community and the desorption of the herbicide to the colloids, thus making it available to the dissipative processes (Blanco, 1979, Walker and Allen, 1984; Reddy and Locke, 1998), reducing its concentration in the soil and thus the beginning of the first plant test germination, the DAT 377 and 475 for treatments 0.6 and 1.2 kg ha^{-1}, respectively (Figure 1).

However, the increase in temperature and precipitation was only sufficient to that the herbicide was dissipated partly because after the dry season (448-556 DAT) was necessary to even more a rainy season (550 until 704 DAT), so that there were further dissipation of the herbicide and progressive reduction of their concentration in the soil, notably for the dose of 0.6 kg ha^{-1}, where it was possible to determine the end of the persistence of sulfentrazone, the highest dose was also detected by this plant test until the end of the evaluations, at 704 DAT.

3.1.3 Soil persistence of sulfentrazone applied in soybean and its effect on crops rotation

Blanco & Velini, 2005 studying the persistence of sulfentrazone applied in soybean cv. Embrapa 48, at the same doses of the test previously described , 0.6 and 1.2 kg ha^{-1}, pH 5.8 and soil organic matter content of 43 g dm-3, using the same bioassay method, also evaluated the effect of this herbicide on five crops in succession: millet cv. Italian and oats cv. White, wheat cv. IAC 24, sunflower cv. Uruguay and kidney bean cv. Carioca, under field conditions for 539 days after the treatments. The biological response of the beet test plant for persistence of the herbicide is described in Figure 7.

The Figure 7 shows the results of the action of sulfentrazone on the beet plant test; it is observed that when the test was carried out with the soybean crop, there were six samples and from these, the growth of beet plants was observed only in the control treatment. After the soybean harvest, it was initiated the preparation of the area for planting crops in rotation, at 159 DAT. Despite all the procedures for preparing the soil for planting, it was found that the management was not sufficient to dissipate the herbicide. The results of the bioassays from samples collected from each culture, at the harvest 278 DAT (oats, beans and millet), 286 DAT (wheat) and 305 DAT (sunflower), demonstrated that the persistence of the herbicide was not influenced by crop type neither by specific cultivation practices that were undertaken for each crop.

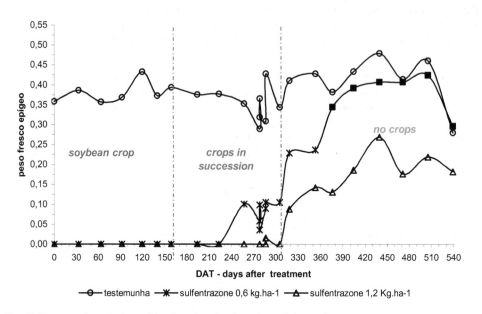

Fig. 7. Temporal variation of fresh epigeal a function of days after treatment, ■ represents no significant difference (p> 0.005), compared to control. Each symbol represents the mean of 5 repetitions (Blanco & Velini, 2005).

In all samplings, tests of means (*t* test at 5% probability) were carried out contrasting the control treatment with herbicide doses in each season. The results indicated that only for the lowest dose, from 376 DAT, there was no significant difference to the control until the end of the test (539 DAT). For this reason, 376 DAT can be defined as the final limit of sulfentrazone persistence at the dose of 0.6 kg ha[-1]. At the highest dose used, the fresh weight of the plant test was significantly lower than that of the control, in all evaluated periods, indicating that, for the dose of 1.2 kg ha[-1], the final limit of persistence was not achieved, thereby; the persistence of sulfentrazone, in soil in this case, was longer than 539 days after herbicide application.

Figure 8 shows the rainfall and average temperature during the research work.

The soil where the assay was carried out showed a high content of organic matter (43 g dm3) and clay (46.3%), this fact combined with the predominant molecular form (85.2%) of the be herbicide, favored the sorption of the herbicide to soil colloids (Weber 1970, Grey et al., 1997). However, as the herbicide was sprayed in the summer, characterized by frequent rainfall and high temperatures (Figure 6), the conditions were not favorable for the herbicide sorption to the soil colloids, but tending to stay in the solution available for dissipative processes and leaching (Briggs, 1976; 1984; Weber, 1970; Walker & Allen, 1984).

This condition was maintained until 80 DAT, the period when due to the dry season, there was the most favorable condition for the herbicide sorption to the colloids, until 200 DAT, when a new rainy condition favored the persistence of the herbicide in the soil solution and subjected to the dissipative processes and leaching. This argument is strengthened by, the

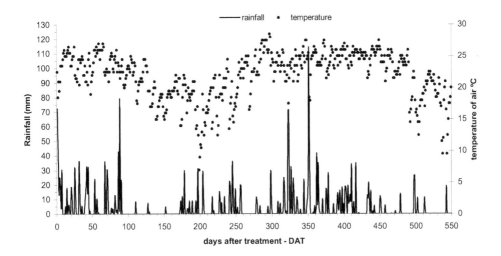

Fig. 8. Daily rainfall and average air temperature in the period from 13/01/2000 to 04/07/2001 (Blanco & Velini, 2005).

first occurrences of test plants in the bioassays at the initial phase of this period, indicating the beginning of the herbicide dissipation.

This condition of high intensity and abundance of rainfall and high temperatures was maintained until almost the end of the assay, at 500 DAT, was favorable to the dissipation of the herbicide, and it was possible to define the limit of the persistence of sulfentrazone in its lowest doses, at 376 DAT. After this period, the weather conditions changed to dry, with cool temperatures, characteristic of the end of the experiment (fall/winter). This phase was favorable to the herbicide sorption to the soil colloids, becoming not available to the processes of dissipation and thus the herbicide residue level in the soil was able to affect significantly the test plant,not being able to detect the final limit of the persistence of sulfentrazone at the highest dose, 1.2 kg ha-1.

For the persistence data obtained in this experiment, beyond the Ecotoxicological aspects, it is important to notice that after the harvest of soybean crop, there were still residues of the herbicide in a reasonable concentration in the soil, resulting in two immediate risks for the crops in succession to soybean: injuries to sensitive crops and when the crop is tolerant to the herbicide, if the same chemical is used for weed control, as it was already present in the soil, there may be an initial concentration higher than the tolerance threshold level and cause damage to the crop that would theoretically be tolerant to the herbicide. To evaluate the selectivity of sulfentrazone on soybean crops in succession, several development parameters of these crops were measured at different periods: visual symptoms of phytotoxicity, stand, height, leaf area, leaf number, fresh and dry weight and also production.

Analysis of the results showed that sulfentrazone, independent of the dose, affected the millet and oat development. On the other hand, for sunflower and bean crops, the variance analysis showed that the treatments were not significant, thus demonstrating the selectivity of this herbicide for these crops. For wheat crop, the selectivity of the herbicide was variable according to the dose tested. It was selective for the lowest dose and not selective for the highest dose, 0.6 and 1.2 kg ha^{-1}, respectively

4. Conclusion

The methodology of bioassays for the detection of residues of sulfentrazone in soil is adequate.

The herbicide when applied in the cultivation of sugarcane cv. SP8018160 and soybean cv. Embrapa 48 shows long persistence in the soil and affects significantly the development of millet cv. Italiano and oats cv. White and wheat cv. IAC 24 (at 1.2 kg ha^{-1}) in soybean rotation, however, for sunflower cv.Uruguay and kidney bean cv . IAC Carioca and wheat cv. IAC 24 (dose 0.6 kg ha^{-1}), sulfentrazone does not affect the development of these plants.

5. References

Alexander, M. (1965). Persistence and biological reaction of pesticides in soil. Soil Science. American Proccedings., v. 29, n 1, p. 1-7

Astolfi, E.; Landoni, J. H.; Almeida, E. (1984). *Curso sobre toxicologia de defensivos agrícolas*, 3. ed. São Paulo. ANDEF, 159p

Blanco, F. M. G (2002). Persistência do herbicida sulfentrazone em solos cultivados com cana-de-açúcar e soja e seu efeito em culturas sucedâneas. Tese de Doutorado em Agronomia Faculdade de Ciências Agrárias, Universidade Estadual Paulista, Botucatu-Brasil

Blanco, F. M. G.; Velini, E. D (2005). Persistência do herbicida sulfentrazone em solo cultivado com soja e seu efeito em culturas sucedâneas. *Planta Daninha*, Vol. 23, No. 4, p. 693-700, 2005, ISSN 0100-8358

Blanco, F. M. G.; Velini, E. D; Batista Filho, A. (2010). Persistência do herbicida sulfentrazone em solo cultivado com cana-de-açúcar. *Bragantia*, Vol. 69, N. 1, p. 71-75, 2010, ISSN 0006-8705

Blanco, H. G (1979). Destino, comportamento e resíduos de herbicidas no solo. *O Biológico*, Vol 45, p. 225-48.

Blanco, H. G.; Oliveira, D. A.; Araújo, J. B. M. (1978). Período crítico de competição de uma comunidade natural de mato em soja (*Glycine max* (L.) Merrill). *O Biologico*, v. 44, p. 299-305

Blanco, H. G.; Oliveira, D. A.; Araújo, J. B. M. (1979). Competição entre plantas daninhas e a cultura da cana-de-açúcar. I. Período crítico de competição produzido por uma comunidade natural de dicotiledôneas em culturas de ano. *O Biológico*, São Paulo, v. 45, p. 131-40

Blanco, H. G.; Oliveira, D. A.; Coleti, J. T. (1981). Competição entre plantas daninhas e a cultura da cana-de-açúcar. II. Período de competição produzido por uma comunidade natural de mato, com predomínio de gramíneas, em culturas de ano. III. Influência da competição na nutrição da cana-de-açúcar. *O Biológico*, São Paulo, v. 47, p. 77-88.

Briggs, G. G. (1976). Degradation in soil. In: Persistence of inseticides and herbicides. The British Crop Council, v. 17, p. 41-54

Briggs, G. G. (1969). Molecular structure of herbicide and their sortion by soil. *Nature,* Vol. 223, No. 288.

Briggs, G. G. (1981) Theoretical and experimental relationships between soil adsorption, octanol-water partition coefficients, water solubilities, bioconcentration factors, and the parachor. *Journal of Agricultural and Food Chemistry,* Vol. 29, p. 1050-1059.

Briggs, G. G. (1984) Factors affecting the uptake of soil-applied chemicals by plants and other organisms., *Proceedings, symposium on soil and crop protection chemicals,* p. 35-47.

EPA - United States Enviromental Protection, Pesticide Fact Sheet (2011) http://www.epa.gov/ opprd001/factsheets/sulfentrazone.pdf

Grey, T. L.; Walker, R. H.; Wehtje, G. R.; Hancock, H. G. (1997). Sulfentrazone adsorption and mobility as affected by soil and pH. *Weed Science,* Vol. 45, p. 733-38.

Harper, S. S. (1994). Sortion-desorption and herbicide behavior in soil. *Rev. Weed Sci.,* v.6, p. 207-25

Hess, D. F. (1993) Herbicide effects on plant structure, physiology, and biochemistry. In: *Pesticide Interactions in Crop Production.* p. 13-34, London: CRC Press Inc, ISBN-13: 9780849363399

Kaufman, D. D.; Kearney, P. C. (1970). Microbial degradation of s-triazine herbicides. Residue Rev., v. 32, p. 235-66,

Madhun, Y. A.; Freed, Y. H. (1990). Impact of pesticides on the environment. In: CHENG, H. H. (Ed.) *Pesticides in the soil environment:* Processes, Impacts, and modeling. Madison: Soil Science of America, p. 429-66

Reddy, K. N. e Locke, M. A. (1998) Sulfentrazone sorption, and mineralization in soil from two tillage systems. *Weed Science,* Vol.. 46, p. 494-500

Rodrigues, B. N.; Almeida, F. S. de (2011). *Guia de herbicidas.* 6a ed. Londrina. ISBN 978-85-905321-2-5

Rossi, C. V. S.; Alves, P. L. C. A.; Marques Junior, J. (2003). Mobilidade do sulfentrazone em nitossolo vermelho e em neossolo quartzarênico. *Planta Daninha,* Vol..21, p.111-120, 2003, ISSN 0100-8358

Seefeldt, S. S.; Jensen, J. E.; Fuerst, E. P. (Log-logistic analysis of herbicide dose-response relationships. *Weed Technology.* vol. 9, p, 218-227

Sheets, T. J. (1970). Persistence of triazine herbicides in soil. *Residue Rev.,* v. 32, p. 287-310

Streibig, J. C. (1988) Herbicide bioassay. Weed Reserch, v. 28, p. 479-484,

Streibig, J. C. et. al. (1995) Variability of bioassay with metsulfuron-metil in soil. Weed Reserch, v. 35, p. 215-24,

Tomlin, C. (1994) *Pesticide Manual.* 10. ed. Cambridge: British Crop Protection Council and The Royal Society of Chemistry, 1341 p. ISBN 0988404795

Velini, E. D. (1992) Comportamento de herbicidas no solo. In: *Simpósio Nacional Sobre Manejo Integrado de Plantas Daninhas em Hortaliças.* FCA-UNESP, Botucatu,

Vivian, R.; Reis, M. R.; Jakelaitis, A.; Silva, A. F.; Guimarães, A. A.; Santos, J. B.; Silva, A. A. (2006) Persistência de sulfentrazone em Argissolo Vermelho-Amarelo cultivado com cana-de-açúcar. *Planta Daninha,* Vol.. 24, p. 741-750, 2006. ISSN 0100-8358

Walker A. e Allen, J. G. (1984) Influence of soil and environmental factors on pesticide. *Soil and Crop Protection Chemistry.,* Vol. 27, p. 27,

Walker, A. (1987) Evaluation of simulation model for prediction of herbicide movement and persistence in soil. *Weed Res.*, Vol. 27, p. 143-152

Walker, A. (1983). The fate and significance of herbicide residue in soil. *Scientific horticulture.* Vol. 34, p. 35-47

Weber, J. B. (1970). Mechanism of adsorption of s-triazines by clay colloids and factors affecting plant availability. *Residue Rev.*, v. 32, p. 93-130

8

Herbicides and the Aquatic Environment

Rafael Grossi Botelho[1], João Pedro Cury[2],
Valdemar Luiz Tornisielo[1] and José Barbosa dos Santos[2]
*[1]Laboratório de Ecotoxicologia, Centro de Energia Nuclear na Agricultura,
Universidade de São Paulo – CENA/USP, Piracicaba, SP,
[2]Universidade Federal dos Vales do Jequitinhonha e Mucuri – UFVJM,
Diamantina, MG,
Brasil*

1. Introduction

The quality of water resources is perhaps currently the most discussed topic when it comes to environmental preservation, since aquatic ecosystems have been suffering changes worldwide in most cases irreversible. Such changes are often associated with human activities such as deforestation, release of industrial and domestic effluents, and even the use of pesticides in agricultural fields, which is one of sources that most contributes to the fall of quality of water resources.

Pesticides are important to the agricultural system. However, it is crucial that they be used with responsibility in order to preserve the quality of the final product and the natural resources that support the production, especially soil and water (Oliveira Junior & Regitano, 2009).

Pesticides are products whose function is to eliminate organisms causing damage to agricultural crops thus ensuring high productivity. Their classification is made according to target species (insecticides, herbicides, fungicides, acaricides, nematicides, etc..) (Alves-Silva & Oliveira, 2003, Sanches et al., 2003), patter of use (defoliants, repellents, and others) (Alves-Silva & Oliveira, 2003; Laws, 1993; Sanches et al., 2003), mechamisns of action (acetylcholinesterase inhibitor, anticoagulants, etc) (Alves-Silva & Oliveira, 2003) or chemical structure (pyretroids, organophosphates, carbamates, etc) (Alves-Silva & Oliveira, 2003; Laws, 1993).

Although these molecules, when applied, have target organisms as their final destination, according to Macedo (2002) 99% of applied pesticides go into the air, water and soil, ie, only 1% reaches its target. This finding is quite disturbing as the world population grows; it means that the use of pesticides will increase (thus increasing food productivity) and natural resources will remain under intense threat from these molecules.

2. Pesticides market in Brazil

Pesticides started to become popular in the middle of the Second World War, when the world discovered the DDT. The ease of accesses of this product and its low cost made it to be extremely used before the discovery of its negative effects. The great successes of this compound in pest control made new products being produced strengthening the agrochemical industry today (Bull & Hathaway, 1986).

Currently, according to the data from National Health Surveillance Agency Anvisa (2010), Brazil is the largest consumer of pesticides in the world and has the largest market for these products with 107 companies authorized to register this compounds, responding for 16% of the world market. According to the sales in Brasil, only in 2010, the industry negotied 342,590 tons of active ingredients and its clear that this number is increasing in recent years (Figure 1).

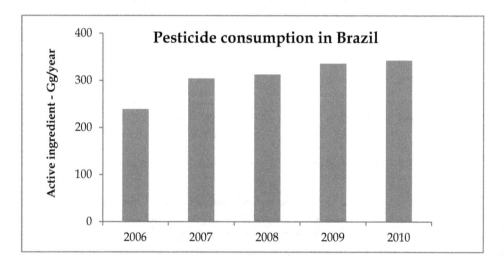

Fig. 1. Pesticide consumption in Brazil, in gig grams of active ingredient, in the period of 2006 to 2010.

Among the classes of pesticides, herbicides are those that make up most marketed worldwide (Moura et al., 2008). These molecules are chemical substances that act by killing or suppressing the development of weeds that impair the productivity of crops of commercial interest (Roman et al., 2007). According to the National Association of Products Industry for Agricultural Defense, only Brazil, one of the leading countries in agriculture with the use of pesticides, 725 000 tons of formulated products were sold in 2009 and herbicides are the main class with 59% (429,693 tons), followed by insecticides and acaricides with 21% (150,189 tons), fungicides 12% (89,889 tons) and others 8% (55,806 tons) (Figure 2) (Sindag, 2010). The problem is that many of these substances are likely to contaminate water resources due to characteristics such as high shift-potential in the soil profile (leaching), high persistence in soil, low to moderate water solubility and moderate adsorption to organic matter present in soil colloids (Almeida et al., 2006). Once present in aquatic environments, these molecules can be absorbed by organisms, and since they live in continual interaction with each other in a complex system of food chains, contamination can result in a drastic imbalance in the ecosystem.

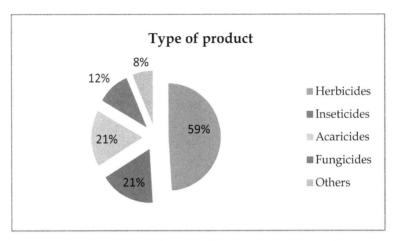

Fig. 2. Pesticide consumption in Brazil by type of product in gig grams of active ingredient, in 2009.

3. Herbicides: leaching and residual effects

Pesticides have an important role in modern agriculture, with new formulations being introduced regularly. Among these, the chlorinated acid-phenoxy herbicides such as 2,4-D and MCPA are commonly used to control weeds in wheat, rice, corn, sugar cane and pasture. The massive use of pesticides has resulted in their presence in the environment in the form of sub-lethal pollution, and problems such as contamination of surface and groundwater have been observed (Legrouri et al., 2005). The concern of environmental protection agencies with the presence of these molecules in soils, water and air has increased greatly in recent times, particularly as it relates to protecting the quality of drinking water (Lagaly, 2001). Due to the commercial importance of agriculture in world and pesticides industry, probably the extensive use of these substances will last for a long period. Therefore, the most feasible would be the rational use of these products through a strict control of its use and handling, aiming, mainly, avoid over dosing, application in undue places and improper washing of packaging and application equipment that many times are held on the banks of rivers (Trovo et al., 2005). Thus, contamination of soils and water due to the extensive use of pesticides over large areas in modern agriculture is a problem that requires research to its remediation (Ignatius et al., 2001).

Considering the transport processes in the environment with which herbicides are related after applied to agricultural areas, leaching and runoff deserve some attention. Surface runoff favors surface water contamination, since the molecule is carried and adsorbed to eroded soil particles or in solution. On the other hand, leaching results in contamination of groundwater, and in this case, chemical substances are carried in solution with the water that feeds the ground water (Spadotto, 2002). Only a low percentage of herbicides in soil are used bioactivity, ie, the remainder is distributed in the environment. This loss of product requires a high amount application, increasing the damage to the environment and consequently to health (Dich et al., 1997).

The knowledge of sorption-desorption processes is of great importance, once determining the amount of product present in the soil is possible to control other processes that can affect the dynamics of these molecules in the soil. If the degree of sorption of a pesticide increases, this compound concentration in water and air decreases. Consequently, the speed of concentration-dependent processes such as volatilization, bioavailability, and vertical movement of pesticides through the soil profile also decrease, thus reducing the risk of contamination of surface and groundwater (Cox et al., 1999).

The aquatic environment has become extremely vulnerable to contamination, Herbicides with high leaching potential, ie, those with low capacity to be retained in the soil are potentially more damaging in this environment by being subject to loading by the underground water flow and deposit with final residual effect on aquatic community. The water pollution is still of concern since often agricultural fields are near lakes, streams and rivers potentiating this environment exposure (Moore et al., 2001) to soluble herbicides. Depend on the physical and chemical characteristics, the residue in water, can bind to the material in suspension, accumulate in the sediment or can be absorbed by aquatic organisms. They can be trasported through the aquatic system by diffusion of water in streans or in the bodies of organisms. Some products may also return to the atmosphere through volatilization. Thus, it is evident that there is a continuous interaction between pesticides, sediment and water, affected by water movement, turbulence and temperature (Nimmo, 1995). This interaction can result in a longer exposure of aquatic organisms to toxic compounds.

Solubility in water is defined as the maximum amount of pure molecule that can be dissolved in water (Lavorenti et al., 2003), being considered the most important physical property related to the transport and fate of organic molecules in aquatic systems, such as herbicides, and also one of determinants of soil sorption coefficient. Thus, herbicides with high solubility have a tendency to be less sorbed to soil colloids (Lavorenti et al., 2003). Therefore, sorption to soil and water solubility becomes important parameters to predict the herbicide trend to move horizontally or vertically in the ground (Extoxnet, 1998).

Another measure of leaching potential for a herbicide is the n-octanol-water partition coefficient (Kow), which measures the hydrophobic or hydrophilic character of a molecule. The Kow is defined as the ratio of the solubility of a compound in octanol (a non-polar solvent) to its solubility in water (a polar solvent). The higher to Kow, the more non-polar the compound (U.S.E.P.A, 2009). In environmental studies, this parameter is also correlated with water solubility, soil sorption coefficient and sediments and bioconcentration in aquatic organisms (Lyman et al., 1990; Sablji et al., 1995; Ran et al., 2002). Herbicides with high log Kow values (> 4.0) or lipophilic tend to accumulate in lipid material, for example, soil organic matter and, consequently, present low mobility (Lavorenti et al., 2003). On the other hand, hydrophilic herbicides (log Kow <1.0) are considered more soluble in water and consequently will present low sorption (Lavorenti et al., 2003) and greater potential for damage to the aquatic community.

Several studies on environmental contamination by pesticides are reported in the literature (Jacomini et al., 2006; Henares et al., 2008). Herbicides of the triazine group, which includes ametryne prometryne, atrazine, simazine, among others, are used worldwide and often detected in samples of soil and water, having high mobility in the environment, lasting years in the ground, water and organisms (Costa Queiroz et al., 1999; Kolpin et al., 2002; Jacomini et al., 2006).

The herbicide atrazine is one of the most widely used herbicides in Brazil, and its use is registred for sorghum, corn, sugarcane and other crops (Rodrigues & Almeida, 2005). Due to its wide use, high persistence and moderate mobility in soil, this herbicide has been detected in several compartments of the environment, especially in surface waters (Buser, 1990) and groundwater (Dörfler et al., 1997). Highest losses of atrazine have been correlated with the first rain or irrigation after its application (Belamie & Gouy, 1992; Patty et al., 1997; Correia et al., 2007). The shorter the time between herbicide application and irrigation or rainfall, the higher the herbicide transport by leaching.

Several authors highlight the problem of contamination of surface and subsurface waters by atrazine (Buser, 1990; Pick et al., 1992; Dörfler et al., 1997; Yassir et al., 1999) so that its use was banned in European Union in 2003 (Sass & Colangelo, 2006). Jablonowski et al. (2009), conducted studies on the persistence of atrazine for more than 20 years after application. Concentrations were detected on average four times higher in the subsurface compared to surface in soil, indicating high risk of contamination of groundwater, even past the experimental period. Armas et al. (2007) found concentrations of atrazine in surface waters of Corumbataí river (São Paulo, Brazil) above the level permitted by Brazilian law. In the United States, atrazine was found at high incidence in surface water and groundwater; research included 178 streams and over 2700 wells (Kolpin et al., 2002). In Australia, atrazine and its metabolites were also detected in low concentrations in groundwater/surface water in several states (Ahmad et al., 2001).

The herbicide clomazone also has high water solubility and persistence in soil and can reach, under aerobic conditions, more than 270 days (California, 2003; Senseman, 2007). After its application on the soil surface, the product may leach into the deeper layers, presenting a potential risk of groundwater contamination and, consequently, watercourses contamination as well (Santos et al., 2008). The clomazone fate and behavior is influenced by organic matter and texture (Loux & Slife, 1989) with edaphic half-life ranging between 5 and 117 days depending on the type of soil and environmental conditions (Curran et al., 1992; Mervosh et al., 1995; Kirksey et al., 1996). Senseman (2007) reports that clomazone persistence is lower in sandy soils than in clay soils.

Monitoring conducted for three years (2000-2003) in two Brazilian rivers (Vacacaí and Vacacaí-Mirim) in Rio Grande do Sul (Brazilian state) detected the presence of the herbicides clomazone, in higher concentration, quinclorac and propanil (Marchesan et al., 2007).

Santos et al. (2008) in a study conducted in shallow waters around the rice-growing areas in Rio Grande do Sul showed that 90% of samples contained clomazone residues. Bortoluzzi et al. (2006) found the presence of this product in surface water adjacent to tobacco crops. The presence of clomazone in water and soil samples have been reported in the literature not only in Brazil but also in other countries such as Spain (Nevado et al., 2007), Italy (Palmisano & Zambonin, 2000), China (Li et al., 2010), Uruguay (Carlomagno et al., 2010) and United States (Gunasekara et al., 2009).

Another herbicide that has concerned researchers is ametryne, whose half-life is between 50 and 120 days in soil and about 200 days in natural water with pH 7.0 and temperature ranging from 5 to 29 °C. This period has been reported as dangerous to the environment by the power of contamination of soils and surface water/groundwater by this product. (Cumming et al., 2002; Laabs et al., 2002; Armas, 2006). Ametryne also has potential to contaminate aquatic environments, once in addition to being transported by runoff, this molecule can undergo leaching. Ametryne residues have also been found in surface waters of Brazil (Armas et al., 2007) although Brazilian law has not yet set a permissible limit in surface waters.

Within the broad class of herbicides, there is no doubt that the most commercialized worldwide is glyphosate. Its occurrence in groundwater was cited only once, in Texas, USA, reported by Hallberg (1989) - under review presented by Amarante Junior et al. (2002) - but the concentration measured was not specified. The direct application as herbicide in surface water to eliminate aquatic plants may be responsible for the presence of glyphosate in surface water.

Due to the rapid adsorption to soil, glyphosate is not readily leached, being unlikely the groundwater contamination. On rare occasions, this herbicide has been detected in water samples, but in general, this occurs due to the difficulty of separating the compounds and also by not being considered a serious water contaminant.

In the case of water pollution, glyphosate can be adsorbed by sediments being carried by them. This interaction is normally fast and occurs within 14 days resulting in much slower natural decay process. The Environmental Protection Agency of the United States (USEPA) sets limits of 700 µg/L glyphosate in drinking water as a "health advisory limit". However, across Europe is established the limit of 0.1 mg/L as "maximum allowable concentration" for pesticides in drinking water as individual substances, since total concentration does not exceed 0.5 mg/L (IAEAC, 1994). Due to its broad-spectrum herbicide properties, ie being non-selective, systemic and low toxicity to animals, as discussed, it became one of the most used in the world, increasing the need for implementation of monitoring programs.

Various processes for water treatment have been investigated regarding their efficiency in removal of certain herbicides present in fresh water samples. Among them, the anaerobic degradation, electrochemical destruction by photo-Fenton reactions, adsorption on activated carbon, adsorption on clays saturated with inorganic or organic cations and the sorption of anionic molecules in lamellar double hydroxides (HDLs) through the processes of anion exchange or merge, among others might be cited.

Atrazine degradation by anaerobic microorganisms was studied by Ghosh & Philip (2004). Authors demonstrated that the degradation of this molecule is dependent on the amount of product in the effluent and the high organic content in the effluent reduces its rate of degradation.

Based on the properties of clays and clay minerals, several authors studied the removal of herbicides present in water, such as phenoxyacetic acid (Yurdakoc & Akcay, 2000), 2,4-D (Hermosín & Cornejo, 1992), prometrine (Socias-Viciana et al., 1998), dicamba (Carrizosa et al., 2001), linuron, atrazine, acephate, diazinon (Villa et al., 1999).

4. Ecotoxicology

The pesticide toxicity is quite complex and overall the goal is to determine what concentration in a particular product is toxic to an organism. The manifestation of a toxic effect resulting from a chemical substance may occur at a point distant from where was found the entry in the medium, since when it reaches a water environment for example, the pollutant can be transported by droplets or particles in suspension thorugh long distances (Pedrozo & Chasin, 2003).

Pollutant toxicity can be expressed by the effective dose or effective concentration (EC_{50} or ED_{50}) which is the amount of a substance affecting half of one group of organisms. Exposure effects to organisms vary according to the product's physico-chemical properties (solubility, chemical reactivity, stability, particle size, etc.) route of exposure (oral, inhalation, dermal), duration and frequency of exposure , species tested (there are differences in susceptibility among species and the type of effect on each one, differences in the effects on individuals of different sex and age, young and elderly are more sensitive than adults), among others (Chasin & Azevedo, 2003). Considering the possibility of contamination in aquatic environments and the need for using herbicides in order to increase agricultural productivity, scientists around the world are working to learn, alert and minimize the effect of these substances in organisms living in these environments. This concern led to the creation of Ecotoxicology, which according to the french toxicologist René Truhaut is the science that studies the effects of natural or synthetic substances on living organisms, populations and communities, animals or plants, terrestrial or aquatic, that make up the biosphere, thus including the interaction of substances with the environment in which organisms live in an integrated context (Plaa, 1982; Niederlehner & Cairns, 1995).

Toxicity tests are used to know the effects of substances in organisms. These represent an important tool in ecotoxicology enabling to determine the toxic effect or not in a particular substance. In the 80's, environmental agencies around the world especially in the United States and Europe began to develop standardized protocols for toxicity test using aquatic organisms (Usepa, 1996; Oecd, 1984-2004). In 1984, the Usepa established the use of organisms for monitoring water quality (USEPA, 1984). Concomitantly, the Organization for Economic Cooperation and Development (OECD) launched a series of test protocols for toxicity to aquatic organisms, including algae, fish and microcrustaceans in Europe. In Brazil, the first initiative to do a focused approach to the subject was in 1975. After this year, other methodologies using groups of organisms have emerged, highlighting algae (Abnt, 1992; Cetesb, 1994), microcrustaceans (Abnt, 1993, 2005; Cetesb, 1994) and fish (Cetesb, 1990).

These tests are used as mechanisms for understanding the effects of anthropogenic impacts on living organisms which act as representative organisms (Campagna, 2005). Toxicity tests allow assessing the environmental contamination by various pollution sources such as agricultural, industrial and domestic waste, chemical products and medicines in general (Marschner, 1999; Lombardi, 2004) and even also detecting the ability of a toxic agent or a mixture to produce deleterious effects showing the extent to which substances are harmful, how and where effects are manifested (Magalhaes & Filho, 2008). They even provide information about the potential danger of a toxic substance to aquatic organisms such as mortality, carcinogenesis, mutagenesis, teratogenesis, behavioral disorders, cumulative physiological, antagonistic and synergistic effects (Baudo, 1987).

The toxicity depends on the susceptibility of the organisms to a particular chemical compound. Different species have different sensitivities according to their feeding habits, behavior, development, physiology and others (Silva & Santos, 2007). Young individuals are usually more susceptible to chemicals than adults, probably due to the diference in degree of development or detoxification mechanisms (Silva & Santos, 2007). Stressed organisms due to of previous exposure to other toxicants may be more sensitive (Rand & Petrocelli, 1985), a commom scenario in the environment .

Toxicity tests are divided into acute and chronic. The acute test aims to assess the effects on organisms to a short period of exposure, whose goal is determining the concentration of a test substance that produces deleterious effects under controlled conditions. For fish, the observed effect is lethality, from which is determined the toxic agent concentration that causes 50% mortality (LC_{50}). For microcrustaceans there is no mobility from which is calculated the average estimate concentration (EC_{50}) that causes 50% immobility (Rand & Petrocelli, 1985). There are also chronic toxicity tests whose organisms are continually exposed to toxic substances for a significant period of time of their life cycle that can vary from half to two thirds of the cycle (Rand & Petrocelli, 1985). These tests assess sublethal effects such as changes in growth and reproduction, changes in behavior (difficulty in movement, increased frequency of opening of the operculum), physiology, biochemistry and tissue changes (Laws, 1993; Adams, 1995). Chronic toxicity tests directly depend on the results of acute toxicity tests, since sublethal concentrations are calculated from the LC_{50} and EC_{50}. For the choice of test organism are often use the following selection criteria: abundance and availability; significant ecological representation within biocenoses; species cosmopolitanism, knowledge of its biology, physiology and dietary habits; genetic stability and uniformity of its populations; low seasonality index, constant and accurate sensitivity; commercial importance; ease of cultivation in the laboratory and, if possible, species should be native for better representation of ecosystems (Rand & Petrocelli, 1995).

Since Ecotoxicology was created several studies have been made always aiming to evaluate the toxicity of a substance for a particular test organism. For example, Botelho et al. (2009) studied the toxicity of various herbicides for tilapia (*Oreochromis niloticus*) including atrazine, paraquat and some mixtures as alaclhor + atrazine, diuron + MSMA and 2,4D + picloram. The LC_{50} (96 hours) was 5.02 mg.L^{-1} for atrazine. These authors also reported weight loss of organisms at 2.5 and 5.0 mg.L^{-1}. In relation to the other products, after 48 hours of exposure, the misture alaclhor + atrazine was the only one that caused 100% of mortality to the organisms. Other studies involving atrazine showed the following LC_{50} values (96 hours): 18.8 mg.L^{-1} for the fish *Cyprinus carpio* (Neskovic et al., 1993), 10.2 mgL^{-1} for *Rhamdia quelen* (Kreutz et al., 2008) and 42.38 mgL^{-1} to *Channa punctatus* (Nwani et al., 2010). In toxicity studies, sensitivity of organisms can vary even if used the same product, as shown in the aforementioned studies with atrazine.

Several other studies involving atrazine has been performed, highlighting Hayes et al. (2002a), which showed that low atrazine concentration (0.1 ppb) stopped the gonadal development of male frogs, confirming the reports of Parshley (2000). In a laboratory study, Hayes et al. (2002b, 2002c) also reported that this herbicide was related to the feminization of *Rana pipiens*. Palma et al. (2009) found that atrazine concentrations affected the reproduction of *Daphnia magna*.

The herbicide atrazine is classified as a toxic agent, carcinogen and hormone disrupter (Friedmann, 2002) which includes potentially carcinogenic compounds to humans (Biradar & Rayburn, 1995). The presence of this product in the environment presents a risk to wildlife and the ecosystem in general, interfering with hormonal activity in animals and human in low doses. Studies have shown that atrazine can also effect the human reproductive system, decreasing the amount of sperms and increasing the infertility (Pan, 2011).

A research of the University of California analyzed the development of 40 males African frogs since the tadpoles stage to adult phase in water with concentration of atrazine within the limits considered safe by the Environmental Protection Agency (EPA). This group of frogs was compared with another without exposure to contaminated water. Among the frogs developed in the water with the herbicide, 10% became functional females. The others 90% despite having characteristics of males, had low testosterone levels and fertility (Hayes, 2010). Strandberg and Scott-Fordsmand (2002) considering organisms exposed to the herbicide simazine, reported ecological effects, including, bioaccumulation in aquatic organisms.

5. Final remarks

There is a growing choice of weed's chemical management by farmers in many agricultural regions of Brazil and the worldwide. The use of herbicides within technical recommendations offers low risk of contamination of non-target sites; however, when applied intensively and without liability, negative environmental impacts may occur.

The application of leachable products such as atrazine and clomazone concerns researchers. It is necessary to achieve sustainable alternatives to the use of these products: the replacement of non-leachable and less toxic products to the environment or follow the banishment example already performed in some countries.

The adoption of bioremediation techniques to areas already contaminated and investment in pesticide application technology as preventive alternative are some of the possibilities to reduce the waste of these molecules in surface water and groundwater.

The lack of supervision by the authorities in small and large agricultural areas coupled with the lack of knowledge of peoples who apply these products and the facility of acquisition contribute to an intensive use and without responsibility.

If professionals and research groups involved in agribusiness can hardly do for the reduction of environmental impacts from domestic and industrial origin, on the other hand, they have fundamental role in the use and dissemination of Good Agricultural Practice to producers, which will be essential for maintaining the activity in a sustainable way at long term.

6. References

Adams, S.M., & Rowland, C.D. (1995). Aquatic toxicology testing methods, In: *Handbook of Toxicology*, Hoffman, D.J., Rattner, B.A., Burton Jr, G.A., & Cairns Jr, J, Boca Raton: Lewis, USA.

Agência Nacional de Vigilância Sanitária (Anvisa). (2010), In: *Agrotóxicos: Agência discute o controle de resíduos no Senado*. 26 de Setembro, Available from: http://www.anvisa.gov.br/divulga/noticias/2009/251109.htm

Ahmad, R.., Kookana, R. S., & Alston, A. M. (2001). Sorption of ametryn and imazethapyr in twenty-five soils from Pakistan and Australia. *Journal of Environmental Science and Health (B)*, Vol. 36, n. 2, pp. 143-160, ISSN 1532-4109.

Akcay, G., & Yurdakoç, K. (2004). Removal of various phenoxy alkanoic acid herbicides from water by organo-clays. *Acta Hydrochimica et Hydrobiologica*, Vol.28, n. 6, pp.300-304, ISSN 0323-4320.

Almeida, S.D.B., Costa, E., Gomes, M.A.F., Luchini, L., Spadotto, C., & Matallo, M. B. Sorção de Triazinas em Solos Tropicais. I. Pré seleção para recomendação de uso na região de Ubatuba, São Paulo, Brasil. In: IV Congreso Iberoamericano de Física Y Química Ambiental, 2006, Cáceres. Medioambiente en Iberoamerica - Visión desde la Física y La Química en los albores del siglo XXI, 2006. Vol. 2. pp. 17-24.

Alves, S.R., & Oliveira-Silva, J.J. É veneno ou e remédio? Agrotóxicos, saúde e ambiente, In: *Avaliação de ambientes contaminados por agrotóxicos*, Peres, F, pp. 137-156, Fiocruz, Rio de Janeiro, Brasil.

Amarante Junior, O. P., Santos, T. C. R., Brito, N. M., & Ribeiro, M. L. (2002). Glifosato: Propriedades, toxicidade, usos e legislação. *Química Nova*, Vol. 25, n. 4, pp. 589-593, ISSN 0100-4042.

Armas, E.D. Biogeodinâmica de herbicidas utilizados em cana-de-açúcar (Saccharum spp) na sub-bacia do rio Corumbataí. 187 p. Tese (Doutorado) – Escola Superior de Agricultura "Luiz de Queiroz", Universidade de São Paulo, Piracicaba, 2006.

Armas, E.D., Monteiro, R.T.R., Antunes, P.M., Santos, M.A.P.F., & Camargo, P.B. (2007). Diagnóstico espaço temporal da ocorrência de herbicidas nas águas superficiais e sedimentos do rio Corumbataí e principais afluentes. *Química Nova*, Vol. 30, n. 5, pp.1119-1127, ISSN 0100-4042.

Associação Brasileira de Normas Técnicas. NBR 12648: Água – Ensaio de toxicidade com *Chlorella vulgaris* (Chlorophyceae). 1992. 8p.

Associação Brasileira De Normas Técnicas. NBR 12713: Água –Ensaio de toxicidade aguda com *Daphnia similis* Claus, 1876 (Cladorcera, Crustacea). 1993. 16p.

Associação Brasileira De Normas Técnicas. NBR 13373: Ecotoxicologia aquática – Toxicidade crônica – Método de ensaio com *Ceriodaphnia* spp (Crustácea, Cladócera). 2005. 15 p.

Baudo, R. Ecotoxicological testing with *Daphinia*. (1987). *Memorie dell'Istituto Italiano di Idrobiologia*, Vol.45, pp.461-482.

Belamie, R., & Gouy, V. (1992). Introduction des pollutants dans le milieu fluvial. Influence du ruissellement des sols. *Oceanis*, Vol.18, pp.505-521.

Biradar, D.P., & Rayburn, A.L. (1995). Chromosomal damage induced by herbicide contamination at concentration observed in public water supplies. *The Science of The Total Environment*, Vol.24, n.6, pp. 1222-1225, ISSN 0048-9697.

Bortoluzzi, E.C., Rheinheimer, D.S., Gonçalves, C.S., Pellegrini, J.B.R., Zanella, R., & Copetti, A.C.C. (2006). Contaminação de águas superficiais por agrotóxicos em função do uso do solo numa microbacia hidrográfica de Agudo, RS. *Revista Brasileira de Engenharia Agrícola e Ambiental*, Vol.10, n.4, pp.881-887, ISSN 1807-1929.

Botelho, R.G.; Santos, J.B.; Oliveira, T.A.; Braga, R.R.; Byrro, E. C. M. (2009). Toxicidade aguda de herbicidas a tilápia (*Oreochromis niloticus*). *Planta Daninha*, Vol 27, n. 3, pp. 621-626, ISSN 0100-8358.

Bull, D., Hathaway, D. (1986). *Pragas e Venenos: Agrotóxicos No Brasil e no Terceiro Mundo*. Petrópolis: Vozes/OXFAM/FASE.

Buser, H.R. (1990). Atrazine and other s-triazine herbicides in lakes and in rain in Switzerland. *Environmental Science and Technology*, Vol. 24, n. 7, pp. 1049-1058, ISSN 0013-936X.

Cairns, J. Jr. (2002). Environmental monitoring for the preservation of global biodiversity: The role in sustainable use of the planet. *International Journal of Sustainable Development and World Ecology*. Vol. 9, n.2, pp. 135-150, ISSN 1745-2627.

California Department of Pesticide Regulation. (2003), In: *Clomazone*. 29 de Agosto de 2011, Available from:
http://www.cdpr.ca.gov/docs/registration/ais/publicreports/3537.pdf.

Campagna, A.F. Toxicidade dos sedimentos da bacia hidrografica do rio Monjolinho (São Carlos – SP): ênfase nas substancias cobre, aldrin e heptacloro. 268 p. Dissertação (Mestrado) – Faculdade de Zootecnia e Engenharia de Alimentos, Universidade de São Paulo, Pirassununga, 2005.

Carlomagno, M., Mathó, C., Cantou, G., Sanborn, J.R., Last, J.A., Hammock, B.D., Roel, A., Carrizosa, M. J., Koskinen, W. C., & Hermosin, M. C. (2001). Dicamba adsorption-desorption on organoclays. *Journal Applied Clay Science*, Vol.18, n.5, pp.223-231, ISSN 0169-1317.

Cavalcanti, D.G.S.M., Martinez, C.B.R., & Sofia, S.H. (2008). Genotoxic effects of Roundup on the fish *Prochilodus lineatus*. *Mutation research*, Vol. 655, n.1-2, pp.41-46, ISSN 1383-5718.

Chasin, A.A.M., & Azevedo, F.A. (2003). Intoxicação e avaliação da toxicidade, In: *As bases toxicológicas da ecotoxicologia*, Azevedo, F.A., & Chasin, A.A.M, pp. 127-165, ISBN 85-86552-64-X, São Paulo.

Chasin, A.A.M., & Pedrozo, M.F.M. (2003). O Estudo da Toxicologia. In: *As bases toxicológicas da ecotoxicologia*, Azevedo, F.A., & Chasin, A.A.M, pp. 1-25, ISBN 85-86552-64-X, São Paulo.

Companhia De Tecnologia De Saneamento Ambiental. Norma CETESB L5.018: Água – Teste de toxicidade aguda com *Daphnia similis* Claus, 1876 (Cladorcera, Crustácea). 1994. 25p.

Companhia De Tecnologia De Saneamento Ambiental. Norma técnica L5.019: Água – Teste de toxicidade aguda com peixes – Parte II – Sistema Semi-Estatico.1990. 29p.

Correia, F.V., Macrae, A., Guilherme, L.R.G., Langenbach, T. (2007). Atrazine sorption and fate in a Ultisol from humid tropical Brazil. *Chemosphere*, Vol. 67, n. 5, pp.847-854, ISSN 0045-6535.

Cox, L., Hermosin, M. C., & Cornejo, J. (1999). Leaching of simazine in organicamended soils. *Communications in Soil Science and Plant Analysis*, Vol.30, n.11-12, pp. 1697-1706, ISSN 1532-2416.

Cumming, J.P., Doyle, R.B., & Brown, P.H. (2002). Clomazone dissipation in four Tasmanian topsoils. *Weed Science*, Vol.50, n.3, pp.405-409, ISSN 1550-2759.

Curran, W. S., Liebl, R.A., & Simmons, F. W. (1992). Effects of tillage and application methods on clomazone, imazaquin, and imazethapyr persistence. *Weed Science*, Vol. 40, n. 3, pp. 482-489, ISSN 1550-2759.

Dich, J., Zahm, S.H., Hanberg, A., & Adami, H.O. (1997). Pesticides and Cancer. *Cancer Causes & Control*, Vol.8, n.3, p. 420-443, ISSN 1573-5243;

Dörfler, U., Feicht, E.A., & Scheunert, I. (1997). S-Triazine residues in groundwater. *Chemosphere*, Vol. 35, n. 1-2, pp. 99-106, ISSN 0045-6535.

Environmental Protection Agency. 12-C-96-114.OPPTS 850.1010 Aquatic invertebrate toxicity test, freshwater daphnids: ecological effects test guidelines. Washington. USA, 1996, 8p.

Environmental Protection Agency. 1984. Technical Support Document for Water Quality-Based Toxic Control. EPA-Washington D.C., 1984. 135 p.

Extoxnet. (1998). Questions about pesticide environmental fate, 29 de Agosto de 2011, Available from: http://extoxnet.orst.edu/faqs/pesticide/pestfate.htm.

Friedmann, A.S. (2002). Atrazine inhibition of testosterone production in rat males following periubertal exposure. *Reproductive Toxicology*, Vol 16, n.3, pp. 275-279, ISSN 0890-6238.

Ghosh, P. K., & Philip, L. (2004). Atrazine degradation in anaerobic environment by a mixed microbial consortium. *Water Research*, Vol.38, n.9, pp. 2277–2284, ISSN 0043-1354.

Glusczak, L., Miron, D.S., Morais, B.S., Simões, R.R., Schetinger, M.R.C., Morsch, V.M., & Loro, V.L. Acute effects of glyphosate herbicide on metabolic and enzymatic parameters of silver catfish (*Rhamdia quelen*). *Comparative Biochemistry and Physiology Part C*, Vol. 146, n.4 , pp.519-524, ISSN 1532-0456.

González, D., & González-Sapienza, G. (2010). Clomazone immunoassay to study the environmental fate of the herbicide in Rice (*Oriza sativa*) agriculture. *Journal of Agricultural and Food Chemistry*, Vol.58, n.7, pp.4367-4371, ISSN 0021-8561.

Gunasekara, A.S., Dela Cruz, I.D., Curtis, M.J., Claasen, V.P., & Tjeerdema, R.S. (2009). The behavior of clomazone in the soil environment. *Pest Mangement Science*, Vol.65, n.6, pp.711-716, ISSN 1526-4998.

Hayes, T.B., Collins, A., Lee, M., Mendoza, M., Noriega, N., Stuart, A.A., & Vonk, A. (2002a). Hermaphroditic, demasculinized frogs after exposure to the herbicide atrazine at low ecologically relevant doses. *Proceedings of the National Academic of Sciences of the United Stades of America*, Vol.99, n.8, pp.5476–5480, ISSN 0027-8424.

Hayes, T.B., Haston, K., Tsui, M., Hoang, A., Haeffele, C., & Vonk, A. (2002b). Atrazine-induced hermaphroditism at 0.1 ppb in American leopard frogs (*Rana pipiens*): laboratory and field evidence. *Environmental Health Perspectives*, Vol. 111, n.4, pp. 568–575, ISSN 0091-6765.

Hayes, T.B., Haston, K., Tsui, M., Hoang, A., Haeffele, C., & Vonk, A. (2002c). Feminization of male frogs in the wild. *Nature*, Vol. 419, pp. 895–896, ISSN 0028-0836.

Hayes, T.B., Khoury, V., Narayan, A., Nazir, M., Park, A., Brown, T., Adame, L., Chan, E., Buchholz, D., Stuave, T., & Gallipeau, S. Atrazine induces complete feminization and chemical castration in male African clawed frogs (*Xenopus laevis*). *Proceedings of The National Academy of Sciences of The United States of America*, Vol. 107, n.10, pp.4612-4617, ISSN 1091-6490.

Henares, M. N. P., Cruz, C., Gomes, G. R., Pitelli, R. A., & Machado, M. R. F. (2008). Toxicidade aguda e efeitos histopatológicos do herbicida diquat na brânquia e no fígado da tilápia nilótica (*Oreochromis niloticus*). *Acta Scientiarum Biological Sciences*, Vol. 30, n. 1, pp. 77-82, ISSN 1807-863X.

Hermosin, M.C., & Cornejo, J. (1992). Removing 2,4-D from water by organo-clays. *Chemosphere*, Vol.24, n. 10, pp.1493-1503, ISSN 0045-6535.

Inácio, J., Taviot-Gueho, C., Forano, C., & Besse, J. P. (2001). Adsorption of MCPA pesticide by MgAl-layered double hydroxides. *Applied Clay Science*, Vol.18, n. 5-6, pp. 255-264, ISSN 0169-1317.

International Association of Environmental Analytical Chemistry - Sample Handling of Pesticides in Water, Active: Barcelona, 1994, p.8-9.

Jablonowski, N.D., Köppchen, S., Hofmann, D., Schäffer, A., & Burauel, P. (2009). Persistence of 14C-labeled atrazine and its residues in a field lysimeter soil after 22 years. *Environmental Pollution*, Vol. 157, n.7, pp. 2126–2131, ISSN 0269-7491.

Jacomini, A.E. Estudo da presença de herbicida ametrina em águas, sedimentos e moluscos, nas bacias hidrográficas do Estado de São Paulo. 113 p. Tese (Doutorado) – Faculdade de Filosofia, Ciências e Letras de Ribeirão Preto, USP, Ribeirão Preto, São Paulo, 2006.

Kirksey, K. B., Hayes, R.M., Krueger, W.A., Mullins, C.A., & Muller, T.C. (1996). Clomazone dissipation in two Tennessee soils. *Weed Science*, Vol. 44, n. 4, pp. 959-963, ISSN 1550-2759.

Kolpin, D.W., Barbash, J. E., & Gilliom, R. J. (2002). Atrazine and Metolachlor ocurrence in shallow ground water of the United States, 1993 to 1995: Relations to explanatarory factors. *Journal of the American Water Resources Association*, Vol.38, n.1, pp. 301-311, ISSN 1752-1688.

Kreutz, L. C., Barcellos, L.J.G., Silva, T.O., Anziliero, D., Martins, D., Lorenson, M., Marteninghe, A., & Silva, L.B. (2008). Acute toxicity test of agricultural pesticides on silver catfish (*Rhamdia quelen*) fingerlings. *Ciencia Rural*, Vol. 38, n. 4, p. 1050-1055, ISSN 0103-8478.

Laabs, V., Amelung, W., Pinto, A.A., Wantzen, M., da Silva, C.J., & Zech, W. (2002). Pesticides in surface water, sediment and rainfall of the northeastern Pantanal basin, Brazil. *Journal of Environmental Quality*, Vol.31, n.5, p.1636-1648, ISSN 0047-2425.

Lagaly, G. (2001). Pesticide–clay interactions and formulations. *Applied Clay Sciense*, Vol, 18, n.5-6, pp.205-209, ISSN 0169-1317.

Lavorenti, A., Prata, F., & Regitano, J. B. (2003). Comportamento de pesticidas em solos: fundamentos. In: *Tópicos em ciência do solo*, Curi, N., Marques, J.J., Guilherme, L.R.G., Lima, J.M., Lopes, A.S., & Alvarez, V.V.H, pp. 335-400, Viçosa: Sociedade Brasileira de Ciência do Solo, ISSN 1519-3934, Viçosa, Brasil.

Laws, E.A. (1993). *Aquatic pollution: an introductory text*. Interscience publication, John Wiley & Sons, ISBN: 978-0-471-34875-7, New York.

Legrouri, A., Lakraimi, M., Barroug, A., De Roy, A., & Besse, J. P. (2005). *Water Research*, Vol. 39, n.15, pp.3441-3448, ISSN 0043-1354.

Li, Y.N., Wu, H.L., Qing, X.D., Li, Q., Li, S.F., Fu, H.Y., Yu, Y.J., & Yu, R.Q. (2010). Quantitative analysis of triazine herbicides in environmental samples by using high performance liquid chromatography and diode array detection combined with second-order calibration based on an alternating penalty trilinear decomposition algorithm. *Analytica Chimica Acta*, Vol.678, n.1, pp.26-33, ISSN 0003-2670.

Lombardi, J.V. (2004). Fundamentos de toxicologia aquática. In: *Sanidade de organismos aquáticos*, Ranzani-Paiva, M.J.T., Takemota, R.M., & Lizama, M.A.P, p.263-272, Livraria Varela, São Paulo.

Loux, M. M., & Slife, F. W. (1989). Availability and persistence of imazaquin, imazethapyr and clomazone in soil. *Weed Sciense*, Vol. 37, n. 2, pp.259-267, ISSN 1550-2759.

Lyman, W. J., Reehl, W. F., & Rosenblatt, D. H. (1990). *Handbook of chemical property estimation methods: Environmental behavior of organic chemicals*, American Chemical Society, ISBN 978-0841217614, Washington.

Macedo, J.A.B. (2002). *Introdução a química ambiental (Química, meio ambiente e sociedade)*, Juiz de Fora, Brasil.

Magalhães, D.P., & Filho, A.S.F. (2008). A ecotoxicologia como ferramenta no biomonitoramento de ecossistemas aquáticos. *Oecologia Brasiliensis*, Vol.12, pp. 355-381, ISSN 1980-6442.

Marchesan, E., Zanella, R., Avila, L. A., Camargo, E. R., Machado, S. L. O., & Macedo, V. R. M. (2007). Rice herbicide monitoring in two brazilian rivers during the rice growing season. *Scientia Agricola*, Vol.64, n.2, pp.131-137, ISSN 0103-9016.

Marschner, A. (1999). Biologische Bodensanierung und ihre Erfolgskontrolle durch Biomonitoring, In: *Okotoxikologie –Okosystemare Ansatze und Methoden Oehlmann*, Oehlmann, J., & Markert, B, pp. 568-576, Ecomed, Landsberg, Alemanha.

Mervosh, T. L., Simms, G. K., & Stoller, E. W. (1995). Clomazone fate as affected by microbial activity, temperature, and soil moisture. *Journal of Agricultural and Food Chemistry*, Vol. 43, n.2, pp. 537-543, ISSN 1520-5118.

Moore, M. T., Rodgers, J.H., Smith, S. Jr., & Cooper, C.M. (2001). Mitigation of metolachlor-associated agricultural runoff using constructed wetlands in Mississippi, USA. *Agriculture, Ecosystems and Environment*, Vol.84, n.2, pp.169-176, ISSN 0167-8809.

Neskovic, N. K., Elezovic, I., Karan, V., Poleksic, V., & Budimir, M. Acute and subacute toxicity of atrazine to carp. *Ecotoxicology and Environmental Safety*, Vol. 25, n.2, pp. 173-182, ISSN 0147-6513.

Nevado, J.J.B., Cabanillas, C.G., Llerena, M.J.V., & Robledo, V.R. (2007). Sensitive SPE CG-MS-SIM screening of endocrine-disrupting herbicides and related degradation products in natural surface waters and robustness study. *Microchemical Journal*, Vol.87, n.1, pp.62-71, ISSN 0026-265X.

Nimmo, D.R. (1985). Pesticides, In: *Fundamentals of aquatic toxicology: methods and applications*, Rand, G.M.., & Petrocelli, S.R, pp 335-373, Hemisphere, ISBN 9780891163022, New York. Organization for Economic Cooperation and Development. OECD Guidelines for Testing Chemicals – Fish,Prolonged Toxicity Test: 14-day Study. Guideline 204. 1984. 9p.

Nwani, C.D., Lakra, W.S., Nagpure, N.S., Kumar, R., Kushwaha, B., & Srivastava, S.K. Toxicity of the Herbicide Atrazine: Effects on Lipid Peroxidation and Activities of Antioxidant Enzymes in the Freshwater Fish *Channa Punctatus* (Bloch). *International Journal of Environmental Research and Public Health*, Vol.7, pp.3298-3312, ISSN 1660-4601.

Oliveira – Junior, R.O., & Regitano, J.B. (2009). Dinâmica de pesticidas do solo, In: *Química e mineralogia do solo*, Alleoni, L.R.F., & Melo, V.F, Viçosa: Minas Gerais, Brasil.

Organization for Economic Cooperation and Development). OECD Guidelines for Testing Chemicals – Fish, Acute Toxicity Test. Guideline 203. 1992. 9p.

Organization for Economic Cooperation and Development). OECD Guidelines for Testing Chemicals – Daphnia magna Reproduction Test. Guideline 211. 1998. 21p.

Organization for Economic Cooperation and Development). OECD Guidelines for Testing Chemicals – Daphnia sp., Acute Immobilization Test. Guideline 202. 2004. 12p.

Palma, P., Palma, V.L., Matos, C., Fernandes, R.M., Bohn, A., Soares, A.M.V.M., & Barbosa, I.R. (2009). Effects of atrazine and endosulfan sulphate on the ecdysteroid system of *Daphnia magna*. *Chemosphere*, Vol.74, n.5, pp.676-681, ISSN 0045-6535.

Pan 2011 – Pesticide Action Network. Atrazine. 29 de Agosto de 2011, available from: http://www.panna.org/resources/specific-pesticides/atrazine#3.

Parshley T. (2000). Report of an Alleged Adverse Effect from Atrazine: Atrazine Technical. EPA Reg. no. 100–529. Washington, DC: U.S. Environmental Protection Agency.

Patty, L., Real, B., & Grill, J.J. (1997). The use of grassed buffer strips to remove pesticides, nitrates and soluble phosphorus compounds from runoff water. *Pesticide Science*, Vol.49, n.3, pp.243-251, ISSN 1096-9063.

Pick, F.E., Van Dyk, L P., & Botha, E. (1992). Atrazine in ground and surface water in maize production areas of the transvaal, South Africa. *Chemosphere*, Vol. 25, n. 3, pp. 335-341, ISSN 0045-6535.

Plaa, G.L. (1982). Present status: toxic substances in the environment. *Canadian Journal of physiology and Pharmacology*, Vol.60, n.7, pp.1010-1016, ISSN 1205-7541.

Queiroz, R.H.C., Lanchote, V.L., Bonato, P.S., Tozzato, E., Carvalho, D., Gomes, M.A., & Cerdeira, A.L. (1999). Determination of ametryn herbicide by bioassay and gas chromatography-mass spectrometry in analysis of residues in drinking water. *Bolletino Chimico Farmaceutico*, Vol.138, n.6, pp.249-252, ISSN 0006-6648.

Ran, Y., He, Y., Yang, G., Johnson, J.L.H., & Yalkowsky, S.H. (2002). Estimation of aqueous solubility of organic compounds by using the general solubility equation. *Chemosphere*, Vol.48, n.5, pp.487–509, ISSN 0045-6535.

Rand. M., & Petrocelli, S.R. (1985). *Fundamentals of aquatic toxicology: Methods and application*. London, Hemisphere Publishing Corporation, USA.

Rodrigues, B.N., & Almeida, F.S. (2005). *Guia de herbicidas*. Grarfmake, ISBN 8590532119, Londrina, Brasil.

Roman, E.E., Beckie, H., Vargas, L., Hall, L., Rizzardi, M.A., & Wolf, T.M. (2007). *Como funcionam os herbicidas da biologia à aplicação*, Gráfica Editora Berthier, Passo Fundo, Brasil.

Sablji, A., Güsten, H., Verhaar, H., & Hermens, J. (1995). QSAR modelling of soil sorption. Improvements and systematics of log KOC vs. log KOW correlations. *Chemosphere*, Vol. 31, n.11-12, pp. 4489–4514, ISSN 0045-6535.

Sanches, S.M., Da Silva, C.H.T., De Campos, S.X., & Vieira, E.M. (2003). Pesticidas e seus respectivos riscos associados a contaminação da água. *Pesticidas: Revista Ecotoxicologia e Meio Ambiente*, Vol.13, n. 0, pp. 53-58, ISSN 1983-9847.

Santos, F.M., Marchesan, E., Machado, S.L.O., Avila, L.A., Zanella, R., & Gonçalves, F.F. (2008). Persistência dos herbicidas imazethapyr e clomazone em lâmina de água do arroz irrigado. *Planta Daninha*, Vol.26, n.4, pp.875-881, ISSN 0100-8358.

Sass, J.B., & Colangelo, A. (2006). European Union bans atrazine, while United States negotiates continued use. *International Journal of Occupational and Environmental Health*, Vol. 12, n. 13, pp. 260-267, ISSN 1077-3525.

Senseman, S. A. (2007). *Herbicide handbook*, Lawrence: Weed Science Society of America.

Silva, J.M., & Santos, J.R. (2007). Toxicologia de agrtóxicos em ambientes aquáticos. *Oecologia Brasiliensis*, Vol.11, n.4, pp.565-573, ISSN 1980-6442.

Sindicato Nacional da Indústria de Produtos para Defesa Agrícola. (2010). 01 de Agosto de 2011, Available from: http://www.sindag.com.br/.

Socias-Viciana, M. M., Hermosin, M. C., & Cornejo, J. (1998). Removing prometrone from water by clays and organic clays. *Chemosphere*, Vol.37, n.2, pp.289-298, ISSN 0045-6535.

Spadotto, C.A. (2002). Indicadores de Impacto Ambiental. Comitê de Meio Ambiente, 14 de Agosto de 2011, Available from: http://www.cnpma.embrapa.br/herbicidas/.

Strandberg, M.T., & Scott-Fordsmand, J.J. (2002). Field effects of simazine at lower trophic levels – a review. *Sciense of the Total Environent*, Vol.296, n.1-3 , pp.117-137, ISSN 0048-9697..

Trovo, A. G., Villa, R. D., & Nogueira, R. F. P. (2005). Utilização de reações foto-Fenton na prevenção de contaminações agrícolas. *Química Nova*, Vol. 28, n.5, pp. 847-851, ISSN 1678-7064.

U.S. Environmental Protection Agency, 2009, Glossary of technical terms: U.S. Environmental Protection Agency, access date May 24, 2011

Villa, M. V., Sanchez-Martin, M. J., Sanchez-Camazano, M. J. (1999). Hydrotalcites and organo-hydrotalcites as sorbents for removing pesticides from water. *Journal of Environmental Science and Health, Part B*, Vol.34, n. 3, pp.509-525, ISSN 1539-4109.

Yassir, A., Lagacherie, B., Houot , S., & Soulas, G. (1999). Microbial aspects of atrazine biodegradation in relation to history of soil treatment. *Pesticide Science*, Vol.55, n. ,pp. 799-809, ISSN 1096-9063.

Zambonin, C.G., & Palmisano, F. (2000). Determination of triazines in soil leachates by solid-phase microextraction coupled to gás chromatography-mass spectrometry. *Journal of Chromatography A*, Vol.874, n.2, pp.247-255, ISSN 0021-9673.

9

Paraquat: An Oxidative Stress Inducer

Ramiro Lascano[1,2], Nacira Muñoz[1,2], Germán Robert[1,2],
Marianela Rodriguez[1], Mariana Melchiorre[1],
Victorio Trippi[1,2] and Gastón Quero[1,2]
[1]*Instituto de Fitopatologia y Fisiologia Vegetal (IFFIVE-INTA), Córdoba,*
[2]*Cátedra de Fisiología Vegetal; Facultad de Ciencias Exactas, Físicas y Naturales; U.N.,*
Córdoba,
Argentina

1. Introduction

Paraquat (1,1_-dimethyl-4,4_-bipyridinium dichloride), is a foliar-applied and non selective bipyridinium herbicides, and it is one of the most widely used herbicides in the world, controlling weeds in a huge variety of crops like corn, rice, soybean, wheat, potatoes; major fruits: apples, oranges, bananas; beverages: coffee, tea, cocoa; and processed crops: cotton, oil palm, sugarcane and rubber.

For a foliar absorbed herbicide to completely kill a plant, it must be capable of accessing the whole plant, as growing leaves and newly emerging roots. This often means that the herbicide not only needs to damage at the point of its absorption, but must also be translocated to parts of the plant not contacted by the herbicide during application.

Paraquat is a cation formed by two pyridine rings, each having a quaternary amine and thus charged 2+. Although the majority of herbicides are passively transported as noionic molecules, paraquat cation movement by diffusion across membrane lipid bilayer is unlikely. Transporter studies to explain paraquat compartment were made using several systems. ABC transporters, large membrane proteins which use ATP for the active transport of several compounds including paraquat have been described. Other groups of transporters are small antiporter proteins which exchange protons for some other molecules using the proton electrochemical potential gradient (Morymio et al., 1992, Yerushalmi, et al., 1995). In animal tissues it has been shown that paraquat transport occurs by carriers that also function as carriers of other molecules such as polyamines (Rannels et al., 1989, Jóri et al., 2007). Hart et al. (1992a 1992b) demonstrated that paraquat movement across plasma membrane root epidermal and cortical maize cells has a concentration-dependent kinetic and that the herbicide binds to cell wall, and its transport is facilitated by a carrier that normally functions in the movement of molecules that has a similar chemical structure or similar charge distribution such us diamines like putrescine and cadaverine. Using maize protoplast Hart et al. (1993) showed that paraquat uptake has similar concentration-kinetic to that observed in intact cells and the accumulation inside cells increase in a time-dependent manner and is saturated after 10 min, although 50% of uptake occurs during the first 10 s. The saturable K_m for paraquat uptake in maize cells and protoplasts was determined at 90 µM and 132 µM respectively, similarly the K_m in rat lung was 70 µM suggesting in both animal and vegetal tissues a carrier-mediated process (Rannels et al., 1985).

In order to investigate paraquat uptake, compartmentation and translocation, maize plantlets with their root immersed in paraquat solution for several loading periods were used (Hart et al., 1993). The lack of chloroplasts in roots provides a system to minimize the short-term phototoxic effect. The paraquat accumulation in the root vacuole was linear over a 24 h loading period. The vacuolar paraquat content, with respect to the total accumulated increased from 15% to 42% after 2 h and 24 h loading period, respectively. In contrast to the vacuole, total cytoplasmic paraquat content appeared to approach saturation whereas paraquat associated with the cell wall fraction remained relatively constant, suggesting that this phase is rapidly saturated. Even though paraquat is considered to be relatively immobile, linear paraquat (PQ) translocation occurred from roots to shoots and was estimated that approximately 50% of the paraquat effluxing from roots started translocation to shoots 5 h after the beginning of loading period (Hart et al., 1993b).

Paraquat acts as a redox cycler with a great negative reduction potential ($E_0 = -0.446$ V). This feature restricts its interaction with strong reductant compounds. When dication of paraquat (PQ^{2+}) accepts an electron from a reductant form the paraquat monocation radical (PQ^+), which then rapidly reacts with oxygen (O_2 $E_0 = 0.16$ V) to initially produce superoxide radical ($O_2^{\bullet-}$) (k 7.7 x 108 M^{-1} s^{-1}) and subsequently the other reactive oxygen species (ROS) such as hydrogen peroxide (H_2O_2) and hydroxyl radical (OH).

In plants, paraquat is principally reduced within chloroplasts, where it acts as an alternative electron acceptor taking electron from Fe-S proteins of photosystem I; inhibiting the ferredoxin reduction, the NADPH generation, and also the regeneration of ascorbic acid. In consequence, paraquat is a potent oxidative stress inducer, because it greatly increases the ROS production and inhibits the regeneration of reducing equivalents and compounds necessary for the activity of the antioxidant system.

Paraquat also induces the increase of superoxide radical production in mitochondria, where complexes I and III are the major electron donors. For this reason paraquat has been widely used to induce mitochondrial oxidative stress in many experimental systems such as isolated mitochondria, cultured cells, and whole organisms including plants, *Saccharomyces cerevisiae*, *Caenorhabditis elegans*, *Drosophila melanogaster* and rodents (Cocheme & Murphy, 2008).

2. Generation and role of ROS

Superoxide radical ($O_2^{\bullet-}$), singlet oxygen ($^1O_2$) hydrogen peroxide (H_2O_2) and hydroxyl radical ($\cdot OH$) are highly reactive compounds that induce protein and pigment degradation, lipid peroxidation, nucleic acid damage, affecting key components of plant cell metabolism that can finally lead to cell death. These deleterious reactions triggered by ROS are known as oxidative stress phenomenon (Casano et al., 1994, 1997; Lascano et al., 1998, 1999).

Even though all ROS are highly reactive compounds their effects and plant responses depend on the ROS in question as well as on its concentration, site of production, interaction with other stress molecules and on the developmental stage and plant cell previous history.

In green tissues under light, chloroplasts are the main intracellular source of ROS (Asada, 1999) and peroxisomes, through photorespiration, are other important ROS producers (del Río et al., 2006). While mitochondria, are the principal source of ROS in darkness and non green tissues. On the other hand, the NADPH oxidase complex, peroxidases and amino oxidases are major sources of apoplastic ROS (Sagi & Fluhr, 2006).

Primarily, the chloroplasts mainly produce $O_2{}^{\cdot-}$ at photosystem I (PSI) and $^1O_2$ at photosystem II (PSII), and the mitochondria produce $O_2{}^{\cdot-}$ at complexes I and III (Asada, 1999). The peroxisomes produce H_2O_2 as byproduct of photorespiratory glycolate oxidase reaction, fatty acid β-oxidation and reaction of flavin oxidase, and $O_2{}^{\cdot-}$ is generated by xanthine oxidase and by electron transport chains in the peroxisomal membrane (del Río et al., 2006).

Various interconverting reactions occur among different ROS. Superoxide is spontaneously or enzymatically converted to H_2O_2 by disproportion mechanism and H_2O_2 and $O_2{}^{\cdot-}$ can interact to produce ·OH through the Fenton reaction catalyzed by free transition metal ions (Fridovich, 1986).

Different ROS have different features. Hydrogen peroxide is a non radical, apolar molecule and, in consequence, it is a relatively stable compound with half-life around 1 ms. In plant tissues, its concentration could be in the micro to millimolar range. The half-lives of the other ROS are very short, ranging from nano to micro second, and then they are present at very low concentrations (Asada, 1999).

Reactive oxygen species also have different reactivities. Hydrogen peroxide (Eo 1.77 V), not a highly reactive ROS *per se*, mainly oxidizes thiol groups, in presence of transition metal ions it catalyzes ·OH generation by Fenton reaction. Superoxide radical (Eo -0.33V) oxidizes ascorbate and NADPH, reduces metal ions and cytochrome C and reacts with protein Fe-S centers. Singlet oxygen is particularly reactive with conjugated double bonds of polyunsaturated fatty acids. Whereas ·OH (Eo 2 V), the most oxidant ROS, reacts with all types of macromolecular cellular components. The differential ROS reactivity means that they leave different footprints in the cell in the form of different oxidatively modified components (Moller et al., 2007).

Cellular membranes are the principal targets of ROS. Peroxidation of polyunsaturated fatty acids (PUFAs) is a common oxidative stress effect. Linoleic acid (18:2) and linolenic acid (18:3) are major fatty acid present in galactolips of thylakois and phospholipids of all membranes. PUFAs peroxidation generates mixtures of lipid hydroperoxides several aldehydes, e.g., 4-hydroxy-2-nonenal (HNE) and malondialdehyde (MDA), hydroxyl and keto fatty acids and oxidative modification in membrane protein. The consequences over the membrane function are the fluidity and selectivity decreases (Halliwell et al.,1999; Halliwell, 2006). Some of the PUFA peroxidation products act directly or after enzymatic modification as secondary messengers either, e.g. oxylipins (Muller et al., 2004).

ROS induce mainly irreversible covalent modification on proteins. The reversible modifications on sulfur containing amino acid are very important in the redox or oxidative signaling. Cystein thiol groups are initially oxidized to disulfide and in further oxidation to sulfenic and sulfinic acid. The highest level of cysteine oxidation, cysteic acid seems to be irreversible and damaging (Ghezzi & Bonetto; 2003). Nitrosylation and glutathionylation are other cystein thiol modification mediated by nitric oxide, reactive nitrogen species (RNS) and glutathione. RNS are generated by the interaction between nitric oxide and ROS. (Costa et al., 2003; Halliwell, 2006). Carbonylation, a common oxidative protein modification affecting particularly Arg, His, Lys, Pro, Thr, and Trp; and conjugation with peroxidation PUFA products, mainly with HNE, are other oxidative protein modifications (Shacter, 2000; Winger et al., 2005).

The generation of 8-Hydroxyguanine is the most common DNA modification induced by ROS. The nucleotide bases are attacked by. OH and $^1O_2$ while H_2O_2 and $O_2^{\cdot-}$ do not react at all (Wiseman & Halliwell, 1996). Chloroplastic and mitochondrial DNAs are into the two major source of ROS where potentially high rates of modification might occur (Thorslund et al., 2002). Another indirect oxidative modification to DNA is the conjugation of MDA with guanine (Jeong, 2005). The DNA oxidative modification could induce changes in cytosines methylation patterns, and then in the regulation of gene expressions. ROS-induced DNA modification seems to be a not completely random process (Halliwell, 2006).

Carbohydrates can be oxidatively modified by ·OH, being the formic acid the main breakdown product of sugar oxidation (Isbell et al., 1973).

In spite of its toxic effects, increasing evidence indicates that ROS are signaling molecules that participate in many processes, such as cell cycle, cell elongation, cell death, plant growth and development, senescence, hormone signaling, responses to biotic and abiotic stress and in symbiotic interaction with microorganisms (Bustos et al., 2008; Mittler et al., 2004; Muñoz et al. 2011, submitted; Rodriguez et al., 2010). The H_2O_2 molecular properties make it a good second messenger that could cross membrane by diffusion or aquaporins. However, all ROS can act as signaling molecules directly or by oxidized product. NADPH oxidase complex, the main source of apoplastic ROS, has a key role in oxidative signaling (Sagi & Fhlur, 2004).

The dual role of ROS, as toxic or signaling molecules, depends on the ratio and subcellular location of its generation, thus the tight regulation of the steady-state level of ROS in different subcellular .compartments has both signaling and oxidative damage protection purposes. The function of ROS as signaling molecules is intrinsically related to the interaction with non-enzymatic antioxidants, such as ascorbate and glutathione, which are redox buffers and also signal molecules *per se* (Foyer & Noctor 2005 a, 2005b).

The relationship among ROS, antioxidants, reducing equivalents, sugars, the redox state of chloroplastic and mitochondria electron transport chains are major determinants of the cellular redox state, which has a critical function in the environmental perception and modulation of defense, acclimation and tolerance responses (Foyer & Noctor, 2005 a; 2005b; Lascano et al., 2003; Melchiorre et al., 2009; Robert et al., 2009).

3. Antioxidant system in plants

Plants have evolved a complex antioxidant system composed by both non-enzymatic and enzymatic components, to prevent the harmful effects of ROS.

Low-molecular-mass metabolites soluble in both aqueous and lipid phases lipid with high ROS reactivity such as ascorbate, glutathione tocopheroles, flavonoids, alkaloids, carotenoids, proline and amines, form non-enzymatic part of the antioxidant system (Apel & Hirt, 2004; Sharma and Dietz, 2006).

Superoxide dismutase (SOD) (E.C.: 1.15.1.1), ascorbate peroxidase (APX) (E.C.: 1.11.1.11), catalase (CAT) (EC 1.11.1.6), and glutathione reductase (GR) (E.C.: 1.6.4.2) are key antioxidant enzymes that modulate the concentration of two of the Haber/Weiss and Fenton reaction substrates, $O_2{}^{\cdot-}$ and H_2O_2, preventing the formation of the highly toxic OH radical (Asada, 1999). Approximately, 80% of SOD, GR, and APX activity is located in the chloroplast (Asada, 1999). CAT activity is located in peroxisomes and mitochondria (Scandalios, 1994). SOD catalyses the disproportionation of $O_2{}^{\cdot-}$ to H_2O_2, and is present in multiple isoforms: copper/zinc (CuZn-SOD), iron (Fe-SOD) and manganese (Mn-SOD) (Bowler et al, 1992). In most plants, CuZn-SOD and Fe-SOD are present in the chloroplasts, CuZn-SOD in the cytosol and Mn-SOD in mitochondria (Casano et al., 1997; Scandalios, 1993). Degradation of H_2O_2, in the chloroplasts and in the cytosol is carried out by the ascorbate-glutathione cycle, which involves APX and GR activities (Lascano et al., 1999, 2003). APX has chloroplastic and cytosolic isoforms, and catalyses the conversion of H_2O_2 to water using ascorbate as electron donor (Asada, 1999).

Reduced glutathione (GSH) and ascorbic acid are the most important soluble non-enzymatic antioxidants and in chloroplasts they are present at millimolar concentrations (Noctor & Foyer, 1998). Ascorbate acts as a ROS quencher and it is involved in the regenerations of tocopherol and violoxanthine depoxidase activity of xanthophylls cycle (Noctor & Foyer, 1998). Reduced glutathione is a tripeptide γ-glutamylcysteinyl glycine (γ-Glu-Cys-Gly) involved in: direct reaction with ROS, the regeneration of the ascorbate pool and as electron donor of glutaredoxins which are linked to type II peroxiredoxin activity. Likewise, GSH participates in the glutathionylation, a post-transcriptional modification of protein thiols groups that regulates the function of proteins like glyceraldehyde-3-phosphate dehydrogenease and thioredoxin activities (Michelet et al., 2005; Zaffagnini et al., 2007). The reduction of oxidized glutathione is NADPH-dependent and carried out by GR, a ubiquitous flavoenzyme with many isoforms, located in chloroplasts, cytosol, and mitochondria (Lascano et al., 2001; Tanaka et al., 1994).

Other more recently identified components of enzymatic antioxidant system are peroxiredoxins and glutathione peroxidase, non-heme-containing peroxidase which activity depend on cystein residues (Bryk et al., 2000; König et al., 2003).

4. The use of paraquat in stress response studies

Plants as sessile organism are permanently exposed to changing environment that become stressful conditions affecting their growth, development and productivity. Tolerance to environmental stress is a major selection criterion in plant breeding. The cellular and molecular tolerance mechanisms of plants to different stresses have been intensively studied.

Reactive oxygen species are produced as byproduct of normal aerobic metabolism and the life under aerobic conditions is strictly dependent on the presence of antioxidant system. Nowadays, it is widely accepted that the generation of ROS is enhanced under abiotic and biotic stress conditions. Depending on stress intensity and its associated-ROS levels the plant responses range from tolerance to death.

Likewise, the positive response of the antioxidant system correlates, in part, with the tolerance to many different environmental stress conditions. ROS and antioxidant system are central components of the cross-tolerance phenomenon which states that a tolerant genotype to one stress condition could be also tolerant to other kinds of stress.

Paraquat treatments have been frequently used, as a potent oxidative stress inducer, in many different basic studies like: oxidative stress tolerance and cross tolerance responses associated with the antioxidant system responses (Lascano et al., 1998, 2001, 2003), forward and reverse genetic approaches to study the function of different antioxidant system components (Melchiorre et al., 2009 and references therein,) ROS signaling (Robert et al., 2009), ROS and NO-induced cell death (Tarantino et al., 2005), and to mimic the drought effect on carbon and nitrogen metabolism of nodules (Marino et al., 2006).

Several attempts made to enhance tolerance photooxidative stress conditions have been tested with paraquat treatments. These have involved the overexpression of enzymes associated with the Asada-Halliwell pathway including SOD (Arisi et al., 1998; Bowler et al, 1991; McKersie et al., 1999; Melchiorre et al., 2009; Perl et al., 1993; Pitcher et al, 1991; Sen Gupta et al., 1993;Tepperman et al, 1990; Van Camp et al.,1996), and GR (Aono et al., 1991, 1993; Creissen et al., 1995; Foyer et al., 1991, 1995; Melchiorre et al., 2009). The tolerance to different oxidative stress conditions was dependent on the copy numbers and overexpression levels; the isoform overexpressed; the subcellular location where the overexpressions were targeted; and the induction of other antioxidant enzymes. The results of chloroplasts-targeted Mn-SOD or GR overexpression in wheat chloroplasts, suggest that antioxidant enzyme overexpression effects on tolerance response not only depend on their antioxidant capacities but also on their effects on the cellular redox state, which modulates the responses to photooxidative stress in a pathway where apoplastic superoxide generation could be involved (Melchiorre et al., 2009). The photooxidative activations of NADPH oxidase complex, the main source of apoplastic ROS, can be mimicked by paraquat treatment (Robert et al., 2009).

Paraquat has also been used as an efficient inducer of cell death in both animal and plant cells (Dodge, 1971; Suntres, 2002). The cell death processes in plants are major regulatory mechanism of growth, development, and responses to biotic and abiotic stresses (Lam et al., 2001; Pennel & Lamb, 1997). Environmental or developmental conditions where cellular redox balance is disturbed and significant ROS accumulation occurred, could lead to the induction of cell death processes (Dat et al., 2000). In this context, two type of ROS-associated stress intensity-dependent death can be defined: Ordered or Programmed Cell Death (PCD) when the cell maintains the membrane and energy generation systems, and Disordered or Necrosis, when these systems are overwhelmed by the oxidative burst. Continuous or transient light-dependent H_2O_2 accumulation, provoke necrosis or PCD, respectively indicating the existence of a ROS levels threshold below which PCD is triggered and above which necrotic cell death prevail (Montillet et al., 2005).

Programmed Cell Death in plant cells shares some similarities with that of animal cells, like organelle degeneration, nuclear condensation, nuclear DNA fragmentation and eventually cell shrinkage. Interestingly, animal anti-apoptotic protein (Bcl-2, Bcl-xL, and CED-9) expressed in plant, prevented apoptosis-like death mediated by chloroplasts photooxidadative stress induced by paraquat (Chen & Dickman, 2004; Mitsuhara et al., 1999).

The in vivo relationship between ROS-associated to environmental stress condition like drought and biological nitrogen fixation (BNF) inhibition in the legume–Rhizobium symbiosis were studied using different dose of paraquat to induce oxidative stress in nodules. Paraquat produced cellular redox imbalance leading to an inhibition of biological nitrogen fixation (BNF). The low paraquat dose provoked BNF decline, preceded by a decrease in sucrose synthase gene expression protein content and activity, while high paraquat induced a faster and more pronounced BNF inhibition, coinciding with a decline in sucrose synthase and also with a reduction in leghaemoglobin content. These results support the occurrence of two regulation pathways for BNF under oxidative stress, one of these involving carbon shortages and the other involving leghaemoglobin / oxygen flux (Marino et al., 2006, 2008).

4.1 Paraquat resistant mutants

To date, several mutants, ecotypes, and biotypes with paraquat resistance have been characterized in a few plant species. Paraquat-resistant mutants have been shown to be cross tolerant to other oxidative stress conditions and have been used to study the tolerance to other photooxidative stress condition (Tsugane et al., 1999).

There are several paraquat-resistant *Arabidopsis* mutants. Photoautotrophic salt tolerate 1 (*pst1*), an *Arabidopsis* mutant that can grow under high salt concentrations, is nearly 10 times more tolerant to paraquat than wild-type seedlings. This mutant, which is also tolerant to high light intensities exhibits higher SOD and APX activities under paraquat, salt, and high light intensities treatments (Tsugane et al., 1999).

The paraquat-resistant *Arabidopsis thaliana* mutant, allelic to the ozone sensitive mutant *rcd1-1*(radical-induced cell death1-1) (Overmyer et al., 2000), called *rcd1-2*, is also tolerant to UV-B and freezing. The tolerance in this mutant is also related to higher levels of the ROS-scavenging enzymes, particularly chloroplastic CuZn-SOD and APX, and also with an increased accumulation of flavonoids (Fujibe et al, 2004). *Arabidopsis* Cvi ecotype also shows a higher resistance to paraquat, which seems to be determined by a new allele of plastidic CuZnSOD (Abarca et al., 2001).

Gigantea, a late-flowering *Arabidopsis* mutant, is resistant to paraquat (Kurepa et al., 1998), however, the resistance mechanism remains unknown (Huq et al., 2000). In the broadleaf weed *Archoteca calendula* (L) paraquat tolerance has been associated with increases in antioxidant defense. This species also exhibit cross tolerance to other stress conditions (Soar et al., 2003).

Arabidopsis paraquat resistant2-1 (par2-1) mutant show an anti-cell death phenotype. Paraquat treatment induce similar superoxide production in *par2-1* and wild-type plants, suggesting that PAR2 acts downstream of superoxide to regulate cell death. *par2-1* encode a S-nitrosoglutathione reductase (GSNOR) that catalyze a major biologically active nitric oxide species, S-nitrosoglutathione. Compared to wild type, *par2-1* mutant showed higher nitric oxide level, suggesting that nitric oxide level and nitrosylation protein modification regulates cell death in plant cells (Chen et al., 2009).

Other paraquat-resistant genotypes have also been reported; like the grass weed *Hordeum glaucum* (Lasat et al., 1997) and *Conyza bonariensis* (Fuerst et al., 1985; Norman et al., 1994). The resistance mechanism seems to be related to a higher herbicide compartmentalization in root vacuoles of the resistant biotype than in the susceptible one. On the contrary, the amount of paraquat accumulated in the cytoplasm of the susceptible biotype was double that found in the resistant biotype.

Additionally, paraquat tolerance has been associated with the expression of transporters able to carry molecules with similar chemical structure or charge distribution to paraquat, like polyamines (Tachihara et al., 2005). Pharmacological treatments with blockers of proton pump ATPases, such us nitrate, carbonyl-cyanide-m-chlorophenylhydrazone (CCCP) and N4N1- dicyclohexylcarbodiimide (DCCD) were used in order to study their effects on paraquat moving into inactive compartments in C. canadiensis (Jóri et al., 2007). Recovery after paraquat treatment in tolerant biotypes was strongly inhibited by nitrate, as nitrate selectively blocks ATPases in the vacuoles -responsible for energy supplies to vacuolar membranes- the results suggested that paraquat sequestration uses energy from the proton gradient (Jóri et al., 2007).

Regarding the relationship between paraquat tolerance and leaf age, some studies have shown that young leaves are more tolerant than mature ones (Kuk et al., 2006; Ohe et al., 2005), it is worth nothing that responses are closely related with detoxify mechanism and antioxidative responses as well as with morphological leaf characteristics such as epicuticular wax content and leaf cuticle development which is the first and most significant barrier for foliar-applied chemicals. Although damage originated by paraquat treatment in *Cucurbita spp* varied among cultivars, the injury provoked by herbicide application was lower in younger leaves than in older ones as it was observed by lesser conductivity values and malondialdehide production which indicate membrane damage with cellular leakage and membrane lipid peroxydation respectively. These responses correlated also with higher antioxidant activity and increases in ascorbate content as well as with higher epicuticular wax in young leaves (Yeol Yoon , 2011).

5. Conclusion

Paraquat is potent oxidative stress inducer, which beyond the widely use as desiccant herbicide, it has been a very useful tool in plant biology basic research. Many aspect of oxidative stress in plants, the toxic and signaling roles of ROS, the native and transgenic plant tolerance/susceptibility responses to many environmental stress conditions, the cross tolerance phenomenon and different cell death processes have been studied using paraquat treatments.

6. References

Abarca, D.; Roldan, M.; Martin, M.; Sabater, B. (2001). *Arabidopsis thaliana* ecotype Cvi shows an increased tolerance to photo-oxidative stress and contains a new chloroplastic copper/zinc superoxide dismutase isoenzyme. *J Exp Bot* 52: 1417–1425.

Aono, M.; Kubo, A.; Saji, H.; Natori, T.; Tanaka, K. and Kondo, N. (1991). Resistance to active oxygen toxicity of transgenic *Nicotiana tabacum* that Expresses the gene for Glutathione reductase from *Escherichia coli*. *Plant Cell Physiol*. 32: 691-697.

Aono, M.; Kubo, A.; Saji, H.; Tanaka, K.; and Kondo, N. (1993). Enhance tolerance to photooxidative stress of transgenic *Nicotina tabacum* with high chloroplastic glutathione reductase activity. *Plant Cell Physiol*. 34: 129-135.

Apel, K. and Hirt, H. (2004). Reactive oxygen species: metabolism, oxidative stress, and signal transduction. *Annu. Rev. Plant Biol*. 55:373–399.

Arisi, A.; Cornic, G.; Jouanin, L.; Foyer, C.H. (1998). Overexpression of iron superoxide dismutase in transformed poplar modifies the regulation of photosynthesis at low

CO_2 partial pressures or following exposure to the prooxidant herbicide methyl viologen. *Plant Physiol.* 117, 565–574

Asada, K. (1999). The water-water cycle in chloroplasts: Scavenging of Active Oxygens and Dissipation of Excess Photons. *Annu Rev Plant Physiol Plant Mol Biol.* 50:601-639

Bowler, C.; Van Montagú, M. Inzé, D. (1992). Superoxide dismutase and stress tolerance. Annu.Rev. Plant Physiol. *Plant Mol. Biol.* 43: 83-116.

Bryk, R.; Griffin, P.; Nathan, C. (2000). Peroxynitrite reductase activity of bacterial peroxiredoxins. *Nature* 407, 211-215.

Bustos, D.; Lascano, R.; Villasuso, A.L; Machado, E.; Racagni, G.; Senn, M.E,; Córdoba, A.; Taleisnik, E. (2008). Reductions in maize root tip elongation by salt and osmotic stress do not correlate with apoplastic $O_2^{\cdot-}$ levels *Annals of Botany* 102: 551-559

Casano, L.M.; Gómez, L. D., Lascano, H.R., González, C. A.; Trippi, V.S. (1997). Inactivation and degradation CuZn-SOD by Active oxygen species in wheat chloroplasts exposed to photooxidative stress *Plant Cell Physiol.* 38: 433-440.

Casano, L.M.; Lascano, H.R.; Trippi, V.S. (1994). Hydroxyl radicals and a thylakoid-bound endopeptidase are involved in the light and oxygen-induced proteolysis in oat chloroplasts *Plant Cell Physiol.* 35:145-15

Chen, S.; Dickman, M. (2004) Bcl-2 Family members localize to tobacco chloroplasts and inhibit programmed cell death induced by chloroplast-targeted herbicides. *J Exp. Bot.* 55 (408) 2617–2623

Chen R.; Sun, S.; Wang, C.;Li, Y.;Liang, Y.;An F.;Li, C.; Dong, H.;Yang X.;Zhang J.; Zuo J. (2009). The *Arabidopsis* PARAQUAT RESISTANT2 gene encodes an S-nitrosoglutathione reductase that is a key regulator of cell death. *Cell Research* 19:1377–1387.

Cochemé, H.M.; Murphy, M.P. (2008). Complex I is the major site of mitochondrial superoxide production by paraquat. *J. Biol. Chem.* 283: (4) 1786- 1798.

Costa, N.J.; Dahm, C.C.; Hurrell, F.; Taylor, E.R.; Murphy, M.P. (2003). Interactions of mitochondrial thiols with Nitric Oxide. *Antioxidant and redox signalling* 5:291-305.

Creissen, G.; Reynolds, H.; Xue, Y.; and Moullineaux, P. (1995). Simultaneuos targeting of pea glutathione reductase and of a bacterial fusion protein to chloroplasts and mitochondria in transgenic tobacco. *The Plant Journal* 8: 167-175.

Dat, J.; Vandenabeele, S.; Vranova,´ E.; Van Montagu, M.; Inze,´ D.; Van Breusegem, F. (2000). Dual action of the active oxygen species during plant stress responses. *Cell Mol Life Sci* 57: 779–795.

del Río, L.; Sandalio, L.; Corpas, F.; Palma, J.; Barroso, J. (2006). Reactive oxygen species and reactive nitrogen species in peroxisomes: Production, scavenging, and role in cell signaling *Plant Physiol.* 141: 330-335.

Dodge, AD. (1971). The mode of action of the bipyridylium herbicides, paraquat and diquat. *Endeavour;* 30:130-135.

Foyer, C.H.; Noctor, G. (2005a). Redox homeostasis and antioxidant signaling: a metabolic interface between stress perception and physiological responses *The Plant Cell* 17:1866-1875.

Foyer, C.H.; Noctor, G. (2005b). Oxidant and antioxidant signalling in plants: a re-evaluation of the concept of oxidative stress. *Plant Cell Environ* 2005; 28:1056-1071.

Foyer, C.H.; Lelandais, M.; Galap, C. and Kunert, K.J. (1991). Effects of elevated cytosolic glutathione reductase activity on the cellular glutathione pool and photosynthesis in leaves under normal and stress conditions. *Plant Physiol.* 97: 863-872.

Foyer, C.H.; Souriau, N.; Lelandais, M.; Kunert, K.J.; Pruvost, C. and Jouanin, L. (1995). Overexpression of glutathione reductase but not glutathione synthetase leads to increases in antioxidant capacity and resistance to photoinhibition in poplar trees. *Plant Physiol.* 109: 1047-1057.

Fridovich, I. (1986). Biological effects of the superoxide radical *Arch. Biochem. Biophys.* 247, I.

Fuerst, E.; Nakatani, H.; Dodge, A.; Penner, D.; Arntzen, C. (1985). Paraquat resistance in Conyza, *Plant Physiol.* 77: 984–989.

Fujibe, T.; Saji, H.; Arakawa, K.; Yabe, N.; Takeuchi, Y.; Yamamoto, K.(2004). A methyl viologen-resistant mutant of *Arabidopsis*, which is allelic to ozone-sensitive rcd1, is tolerant to supplemental ultraviolet-B irradiation. *Plant Physiol.* 134:275-285.

Halliwell B. (2006). Reactive species and antioxidants. redox biology is a fundamental theme of aerobic life. *Plant Physiol.* 141:312-322

Hart, H.; Di Tomaso, J.; Kochian, L.(1993). Characterization of paraquat transport in protoplasts from maize (*Zea mays* L.) suspension cells *Plant Physiol.* 103: 963-969.

Hart, J.J.; DiTomaso, J.M.; Linscott, D.L.; Kochian, L.V. (1992a). Characterization of the transport and cellular compartmentation of paraquat in roots of intact maize seedlings. *Pestic Biochem Physiol* 43: 212-222.

Hart, J.J.; DiTomaso, J.M.; Linscott, D.L.; Kochian, L.V. (1993). Investigations into the cation specificity and metabolic requirements for Paraquat transport in roots of intact maize seedlings *Pestic Biochem Physiol* 45: 62-71.

Hart, J.J.; DiTomaso, J.M.; Linscott, D.L.; Kochian, L.V.(1992b). Transport interactions between paraquat and polyamines in roots of intact maize seedlings. *Plant Physiol.* 99:1400-1405.

Huq, E., Tepperman, J.M., and Quail, P.H. (2000). GIGANTEA is a nuclear protein involved in phytochrome signaling in *Arabidopsis*. *PNAS* 97: 9789–9794.

Isbell, H.S.; Frush, H.L.; Martin, E.T. (1973). Reactions of carbohydrates with hydroperoxides. Reactions of carbohydrates with hydroperoxides: Part I. Oxidation of aldoses with sodium peroxide. *Carbohydrate Research* 26:287-295.

Jeong, Y.C.; Nakamura, J.; Upton, P.B.; Swenberg, J.A.(2005). Pyrimido[1,2-α]-purin-10(3H)-one, M1G, is less prone to artifact than base oxidation. *Nucleic Acids Res.* 33:6426 - 6434.

Jóri, B.; Soós,V.; Szego, D.; Páldi, E.; Szigeti, Z.; Rácz, I.; Lásztity, D. (2007). Role of transporters in paraquat resistance of horseweed Conyza canadensis (L.) Cronq. *Pestic. Biochem. Physiol.* 88 :57–65.

König, J.; Lotte, K.; Plessow, R.; Brockhinke, A.; Baier, M.; Dietz, K.J. (2003). Reaction mechanism of plant 2-Cys peroxiredoxin: Role of the C terminus and the quaternary structure *J. Biol. Chem.* 278: 24409-24420.

Kuk, Y.; Shin, J.;Jung, H.; Guh, J.; Jung, S.; Burgos, N. (2006). Mechanism of tolerance to paraquat in cucumber leaves of various ages, *Weed Sci.* 54: 6–15.

Kurepa, J.;Smalle, J.; Van Montagu, M.; Inzé, D. (1998). Oxidative stress tolerance and longevity in *Arabidopsis*: the late flowering mutant gigantea is tolerant to paraquat. *The Plant Journal*, 14: 759-764.

Lam, E.; Kato, N.; Lawton, M. (2001). Programmed cell death mitochondria and the plant hypersensitive response *Nature* 411:826-833.

Lasat, M.; DiTomaso, J.; Hart, J.; Kochian, L. (1997). Evidence for vacuolar sequestration of paraquat in roots of a paraquat-resistant Hordeum glaucum biotype *Physiol Plantarum* 99: 255–262.

Lascano, H.R.; Antonicelli, G.E.; Luna, C.M.; Melchiorre, M.N.; Racca, R.W.; Trippi, V.S. Casano, L.M. (2001). Antioxidative system response of different wheat cultivars under drought: Field and in vitro studies Australian Journal of *Plant Physiology* 28:1095-1102.

Lascano, H.R.; Gómez, L.D.; Casano L.M.; Trippi, V.S (1999). Wheat chloroplastic glutathione reductase activity is regulated by the combined effect of pH, NADPH and GSSG *Plant Cell Physiol.* 40: 683-690.

Lascano, H.R.; Gómez, LD.; Casano, L.M.; Trippi V.S (1998). Changes in glutathione reductase activity and protein content in wheat leaves and chloroplasts exposed to photooxidative stress *Plant Physiol. Biochem.* 36: 321-329.

Lascano, H.R.; Melchiorre, M.N.; Luna, C.M.; Trippi, V.S. (2003). Effect of photooxidative stress induce by paraquat in two wheat cultivars with differential tolerance to water stress *Plant Science* 164: 841-846.

Lorrain, S.; Vailleau, F.; Balague,´ C.; Roby, D. (2003). Lesion mimic mutants: keys for deciphering cell death and defense pathways in plants? *Trends Plant Sci* 8: 263–271.

Marino, D.; Gonzalez, E.; Arrese-Igor, C. (2006). Drought effects on carbon and nitrogen metabolism of pea nodules can be mimicked by paraquat: evidence for the occurrence of two regulation pathways under oxidative stresses *J. Exp. Bot* .57 (3): 665–673.

Marino, D.; Hohnjec, N.; Küster, H.; Moran, J. González, E.; Arrese-Igor, C. (2008). Evidence for transcriptional and post-translational regulation of sucrose synthase in pea nodules by the cellular redox state *MPMI* 21 (5): 622–630.

McKersie, B.D.; Bowley, S.R.; Jones, K.S. (1999). Winter survival of transgenic alfalfa overexpressing superoxide dismutase. *Plant Physiol.* 119: 839–848.

McKersie, B.D.; Chen, Y.; de Beus, M.; Bowley, S.R.; Bowler, C.; Inzé, D.; D`Halluin, K.; Botterman, J., (1993). Superoxide dismutase enhances tolerance of freezing stress in transgenic alfalfa (Medicago sativa L.). *Plant Physiol.* 103:1155-1163.

Melchiorre, M; Robert, G.; Trippi, V.; Racca, R.; Lascano HR (2009). Superoxide Dismutase and Glutathione Reductase overexpression in wheat protoplast: changes in cellular redox state and photooxidative stress tolerance. *Plant and Growth Regulation* 57 (1): 57-68.

Michelet, L.; Zaffagnini, M.; Marchand. C,; Collin, V.; Decottignies, P.; Tsan, P.; Lancelin, J.M.; Trost, P.; Miginiac-Maslow, M.; Noctor, G.; Lemaire, S. (2005). Glutathionylation of chloroplast thioredoxin f is a redox signaling mechanism in plants *PNAS* 102(45):16478-16483.

Mitsuhara, I.; Malik, K.; Miura, M.; Ohashi, Y. (1999). Animal cell-death suppressors Bcl-xL and Ced-9 inhibit cell death in tobacco plants. *Current Biology* 9:775–778.

Mittler, R.; Vanderauwera, S.; Gollery, M.; Van Breusegem, F. (2004). The reactive oxygen gene network of plants. *Trends Plant Sci.* 9, 490-498.

Mittler. R.; Rizhsky, L. (2000) Transgene-induced lesion mimic. Plant Mol Biol 44: 335–344

Moller, I.; Jensen, P.; Hansson, A. (2007). Oxydative modifications to cellular components in plants. *Ann Rev Plant Biol*. 58: 459-481.

Montillet, J.L; Chamnongpol, S.; Ruste'rucci, C.; Dat, J.; Van de Cotte, B.; Agnel, J.P.; Battesti, C.; Inze,' D.; Van Breusegem, F.; Triantaphylide`s, C. (2005). Fatty acid hydroperoxides and H2O2 in the execution of hypersensitive cell death in tobacco leaves. *Plant Physiol* 138: 1516-1526.

Morimyo, M.; Hongo. E.; Hama-Inaba, H.; Machida, I. (1992). Cloning and characterization of the mvrC gene of Escherichia coli K-12 which confers resistance against methyl viologen toxicity, *Nucleic Acid Res*. 20:3159-3165.

Mueller, M.J. (2004). Archetype signals in plants: The phytoprostanes. *Curr. Opin. Plant Biol*. 7: 441-448.

Muñoz, N.; Robert, G.; Melchiorre, M.; Racca, R.; Lascano, R.. (2011). Saline and osmotic stress differentially affects apoplastic and intracellular reactive oxygen species production, curling and death of root hair during Glycine max L.-Bradyrhizobium japonicum interaction Environmental and experimental botany (submitted)

Noctor, G.; Foyer, C.H. (1998). Ascorbate and glutathione. Keeping active oxygen under control. *Annu. Rev. Physiol. Plant Mol. Biol*. 49:249-279.

Norman, M.; Smeda, R.; Vaughn, K.; Fuerst, E. (1994) Differential movement of paraquat in resistant and sensitive biotypes of Conyza. *Pestic. Biochem. Physiol*. 50 : 31-42

Ohe, M.; Rapolu, M.; Mieda, T.; Miyagawa, Y.; Yabuta, Y.; Yoshimura, K.; Shigeoka, S. (2005). Decline in leaf photooxidative-stress tolerance with age in tobacco, *Plant Sci*. 168: 1487-1493.

Overexpression of glutathione reductase but not glutathione synthetase leads to increases in antioxidant capacity and resistance to photoinhibition in poplar trees. *Plant Physiol*. 109: 1047-1057.

Overmyer, K.; Tuominen, H.; Kettunen, R.; Betz, C.; Langebartels, C.; Sandermann, H. Jr; Kangasjärvi, J. (2000). Ozone-sensitive *Arabidopsis* rcd1 mutant reveals opposite roles for ethylene and jasmonate signaling pathways in regulating superoxide-dependent cell death. *Plant Cell* 12: 1849-1862.

Pennel, R.; Lamb, C. (1997). Programmed cell death in plants. *Plant Cell* 9:1157-1168.

Perl A, Perl-Treves R, Galli S, Aviv D, Shalgi E, Malkin S, Galun E. (1993). Enhanced oxidative-stress defense in transgenic potato expressing tomato Cu/Zn superoxide dismutases. *Theor. Appl. Genet*. 85:568-576.

Pitcher, L.H.; Brennan, E.; Hurley, A.; Dunsmuir P.; Tepperman, J.M.; Zilinskas, B. (1991). Overproduction of Petunia chloroplastic copper/zinc superoxide dismutase does not confer ozone tolerance in transgenic tobacco. *Plant Physiol*. 97: 452-455.

Rannels, D.E.; Kameji, R.; Pegg, A.E,; Rannels, S.R. (1989). Spermidine uptake by type I1 pneumocytes: interactions of amine uptake pathways. *Am J Physiol* 257: L346-L353

Rannels, D.E; Pegg, A.E.; Clark, R.S.; Addison, J.L. (1985). Interaction of paraquat and amine uptake by rat lungs perfused in situ. *Am. J. Physiol* 249: E506-E513.

Robert, G; Melchiorre, M.; Trippi, V.; Lascano H.R. (2009). Apoplastic superoxide generation in wheat protoplast is regulated by chloroplastic ROS generation: effects on the antioxidant system *Plant Science* 177:168-174.

Rodríguez, M.; Taleisnik, E.; Lenardon, S.; Lascano R. (2010). Are Sunflower chlorotic mottle virus infection symptoms modulated by early increases in leaf sugar concentration? *J Plant Physiol*. 167(14):1137-1144.

Sagi, M.; Fluh, R. (2006). Production of Reactive Oxygen Species by Plant NADPH Oxidases *Plant Physiol* 141:336-340.

Scandalios, J.G. (1994). Regulation and properties of plant catalases. *In Causes of photooxidative stress and amellioration of defense systems in plants.* Foyer C.; Mullineaux, P. (Eds.) 275-315. CRC Press, Boca Raton, Florida, USA.

Sen Gupta, A.; Webb, R.; Holaday, A.; Allen, R. (1993). Overexpression of superoxide dismutase protects plants from oxidative stress. Induction of ascorbate peroxidase in superoxide dismutase-overexpressing plants. *Plant Physiol.* 103:1067-1073.

Shacter, E. (2000). Quantification and significance of protein oxidation in biological samples. *Drug Metabolism Reviews* 32: 307-326.

Sharma, S.; Dietz, K.J. (2006). The significance of amino acids and amino acid-derived molecules in plant responses and adaptation to heavy metal stress *J. Exp. Bot.* 57 (4): 711-726.

Soar, C.; Karotam, J.; Preston, C.; Ponles R. (2003). Reduced paraquat translocation in paraquat resistant Arctotheca calendula (L.) Levyns is a consequence of the primary resistance mechanism, not the cause *Pesticide Biochem. Physiol.* 76:91–98.

Suntres, Z.E. (2002). Role of antioxidants in paraquat toxicity. *Toxicology* 180:65-77.

Tachihara, K.; Uemura, T.; Kashiwagi, K.; Igarashi, K. (2005). Excretion of putrescine and spermidine by the protein encoded by YKL174c (TPO5) in Saccharomyces cerevisiae, *J. Biol. Chem.* 280: 12637– 12642.

Tanaka, K.; Sano, T.; Ishizuka, T.; Kitta, K.; Kamura, Y. (1994). Comparison of properties of leaf and root glutathione reductases from spinach. *Physiol. Plant.* 91: 353-358.

Tarantino, D.; Vannini, C.; Bracale, M.; Campa, M.; Soave, C.; Murgia, I. (2005). Antisense reduction of thylakoidal ascorbate peroxidase in *Arabidopsis* enhances Paraquat induced photooxidative stress and nitric oxide-induced cell death. *Planta* 221: 757–765.

Tepperman, J.M.; Dunsmuir, P. (1990). Transformed plants with elevated levels of chloroplastic SOD are not more resistant to superoxide toxicity. *Plant Mol. Biol.*14: 501-511.

Thorslund, T.; Sunesen, M.; Bohr, V.A.; Stevnsner, T. (2002). Repair of 8-oxoG is slower in endogenous nuclear genes than in mitochondrial DNA and is without strand bias. *DNA Repair* 1: 261-276.

Tsugane, K.; Kobayashi, K.; Niwa, Y.; Ohba, Y.; Wada, K.; Kobayashi, H. (1999). A recessive *Arabidopsis* mutant that grows photoautotrophically under salt stress shows enhanced active oxygen detoxification *Plant Cell* 11: 1195-1206.

Van Camp, W.; Capiau, K.; Van Montagu, M.; Inzé, D.; Slooten, L. (1996). Enhancement of oxidative stress tolerance in transgenic tobacco plants overproducing Fe-superoxide dismutase in chloroplasts. *Plant Physiol.* 112:1703–1714.

Winger, A.M.; Millar, AH.; Day, D.A. (2005). Sensitivity of plant mitochondrial terminal oxidases to the lipid peroxidation product 4-hydroxy-2-nonenal (HNE) *Biochem. J.* 387: 865–870.

Wiseman, H.; Halliwell, B. (1996). Damage to DNA by reactive oxygen and nitrogen species: role in inflammatory disease and progression to cancer. *Biochem. J.* 313:17-29.

Yeol Yoon, J.; San Shin, J.; Young Shin, D.; Hwan Hyun, K., Burgos, N.; Lee, S.; Kuk, Y. (2011). Tolerance to paraquat-mediated oxidative and environmental stresses in squash (Cucurbita spp.) leaves of various ages. *Pesticide Biochem. Physiol.* 99: 65–76.

Yerushalmi H, Lebendiker M, Shuldiner S. (1995). EmrE, an Escherichia coli 12-kDa multidrug transporter, exchanges toxic cations and H+ and is soluble in organic solvents, *J. Biol. Chem.* 270: 6856–6863

Zaffagnini, M.; Michelet, L.; Marchand, C.; Sparla, F.; Decottignies, P.; Le Maréchal, P.; Miginiac-Maslow, M.; Noctor, G.; Trost, P.; Lemaire, S. (2007). The thioredoxin-independent isoform of chloroplastic glyceraldehyde-3-phosphate dehydrogenase is selectively regulated by glutathionylation *FEBS Journal* 274 (1):212-226.

10

Comparative Assessment of the Photocatalytic Efficiency of TiO$_2$ Wackherr in the Removal of Clopyralid from Various Types of Water

Biljana Abramović[1], Vesna Despotović[1], Daniela Šojić[1], Ljiljana Rajić[1],
Dejan Orčić[1] and Dragana Četojević-Simin[2]
*[1]Faculty of Sciences, Department of Chemistry,
Biochemistry and Environmental Protection, Novi Sad,
[2]Oncology Institute of Vojvodina, Sremska Kamenica,
Serbia*

1. Introduction

Many pyridine derivatives have found widespread application as herbicides. Because of their frequent use, chemical stability and resistance to biodegradation, they are encountered in waste waters, and, due to their hazardous effects on ecosystems and human health, their removal is imperative (Stapleton et al., 2006). With this in mind, we have recently paid significant attention to the study of the model compounds (Abramović et al., 2003; Abramović et al., 2004a, 2004b) and pyridine containing pesticides (Abramović et al., 2007; Šojić et al., 2009; Abramović & Šojić, 2010; Abramović et al., 2010; Guzsvány et al., 2010; Šojić et al., 2010a, 2010b; Banić et al., 2011).

Clopyralid (3,6-dichloro-2-pyridinecarboxylic acid, CAS No. 1702-17-6, $C_6H_3Cl_2NO_2$, M = 192.00 g mol^{-1}) (CLP) is a systemic herbicide from the chemical class of pyridine compounds, i.e., pesticides of picolinic acid. It has been used effectively for controlling annual and perennial broadleaf weeds in certain crops and turf. It also provides effective control of certain brush species on rangeland and pastures. The acidic form of CLP and three CLP salts (triethylamine, triisopropylamine, and monoethanolamine), which are very soluble in water, are commonly used in commercial herbicide products. Its chemical stability along with its mobility allows this herbicide to penetrate through the soil, causing long-term contamination of the ground water, as well as surface water supplies (Cox, 1998; Huang et al., 2004; Donald et al., 2007; Sakaliene et al., 2009). Due to these properties, CLP has recently been reported to occur in drinking water at concentrations above the Permitted Concentration Value of 0.1 µg L^{-1} for an individual pesticide (EU directive 98/83/EC). Although the occurrence of CLP in surface, ground and drinking waters has been widely reported, there are only a few studies concerning with its photocatalytic removal from water (Šojić et al., 2009; Šojić et al., 2010a, 2010b; Tizaoui et al., 2011). These studies showed that the degradation of this herbicide takes place most effectively in the presence of Degussa P25 as photocatalyst. However, several recent studies of photocatalytic activity reported that some cosmetic pigments (TiO$_2$, Wackherr's

"Oxyde de titane standard") are even more efficient than TiO_2 Degussa P25 in the photodegradation of phenol (Rossatto et al., 2003; Vione et al., 2005) and herbicides with a pyridine ring (Abramović et al., 2011).

The aim of this work was to study the effect of water type (double distilled (DDW), tap and river water) on the efficiency of TiO_2 Wackherr toward photocatalytic degradation of CLP. First of all, the study is concerned with the transformation kinetics and efficiency of photocatalytic degradation of CLP in DDW. The study encompasses the effects of a variety of experimental conditions such as the effect of the type of irradiation, catalyst loading, the initial concentration of CLP, temperature, pH, presence of electron acceptors, and hydroxyl radical ($^•$OH) scavenger on the photodegradation kinetics in DDW. The results were compared to the most often used TiO_2 Degussa P25. An attempt has also been made to identify the reaction intermediates formed during the photo-oxidation process of CLP, using the LC–ESI–MS/MS method. The cell growth activity of CLP alone or in the mixture with its photocatalytic degradation intermediates was evaluated *in vitro* in rat hepatoma and human fetal lung cell line, using colorimetric Sulphorhodamine B assay. Finally, the matrix effect of river and tap water on photocatalytic removal of CLP was also studied.

2. Experimental

2.1 Water samples, chemicals and solutions

All chemicals were of reagent grade and were used without further purification. CLP, 99.4%, pestanal quality, was manufactured by Riedel-de Haën; 85% H_3PO_4 was obtained from Lachema (Neratovice, Czech Republic) and NaOH from ZorkaPharm (Šabac, Serbia). The other chemicals used, such as 30% H_2O_2, *cc* acetic acid and 96% ethanol, were obtained from Centrohem (Stara Pazova, Serbia), $KBrO_3$, $(NH_4)_2S_2O_8$ and 60% $HClO_4$, from Merck, while 99.8% acetonitrile (ACN) and HPLC gradient grade methanol (MeOH) were products of J. T. Baker and humic acids (HUM) technical, of Fluka. All solutions were made using DDW. The pH of the reaction mixture was adjusted using a dilute aqueous solution of $HClO_4$ or NaOH. Aspirin® was purchased from Bayer, doxorubicin (Doxorubicin-Teva®) from Pharmachemie B.V. (Haarlem, Netherlands) and gemcitabine (Gemzar®) from Lilly France S.A. (Fegersheim, France), fetal calf serum (FCS) from PAA Laboratories GmbH (Pashing, Austria), penicillin and streptomycin from Galenika (Belgrade, Serbia), trypsin from Serva (Heidelberg, Germany), and EDTA, trichloroacetic acid (TCA), mercury(II) chloride from Laphoma (Skopje), and tris(hydroxymethyl)amino methane (TRIS) from Sigma Aldrich.

Wackherr's "Oxyde de titane standard" (100% anatase form, surface area 8.5 ± 1.0 m^2 g^{-1}, crystallite size 300 nm (Vione et al., 2005) hereafter "TiO_2 Wackherr", and TiO_2 Degussa P25 (75% anatase and 25% rutile form, 50 m^2 g^{-1}, about 20 nm, non-porous) were used as photocatalysts.

The tap water sample was taken from the local water supply network (Novi Sad, Serbia). River water, collected from the Danube (Novi Sad, Serbia) in May 2010, was filtered through Whatman filter paper 42 (diameter: 125 mm, pore size: 0.1 μm, ashless) before use. The physicochemical characteristics of the water samples, along with that of DDW are given in Table 1.

Parameter	Water type		
	DDW	Tap water	Danube river
pH	6.5	7.3	7.8
El. conductivity at 25 °C (μS mL^{-1})	2.9	516	365
TOC (mg L^{-1})	1.04	1.80	5.60
Carbonate hardness (°dH)	0.37	13.06	8.36
HCO$_3^-$ (mg L^{-1})		285	182

Table 1. The physicochemical characteristics of the analysed water types.

2.2 Photodegradation procedures

The photocatalytic degradation was carried out in a cell made of Pyrex glass (total volume of ca. 40 mL, liquid layer thickness 35 mm), with a plain window on which the light beam was focused. The cell was equipped with a magnetic stirring bar and a water circulating jacket. A 125 W high-pressure mercury lamp (Philips, HPL-N, emission bands in the UV region at 304, 314, 335 and 366 nm, with maximum emission at 366 nm), together with an appropriate concave mirror, was used as the radiation source. Irradiation in the visible spectral range was performed using a 50 W halogen lamp (Philips) and a 400 nm cut-off filter. The outputs for the mercury and halogen lamps were calculated to be ca. 8.8×10^{-9} Einstein mL^{-1} min^{-1} and 1.7×10^{-9} Einstein mL^{-1} min^{-1} (potassium ferrioxalate actinometry), respectively. In a typical experiment, and unless otherwise stated, the initial CLP concentrations were 1.0 mM, and the TiO$_2$ Wackherr loading was 2.0 mg mL^{-1}. The total suspension volume was 20 mL. The aqueous suspension of TiO$_2$ Wackherr was sonicated (50 Hz) in the dark for 15 min before illumination, to uniformly disperse the photocatalyst particles and attain adsorption equilibrium. The suspension thus obtained was thermostated and then irradiated at a constant stream of O$_2$ (3.0 mL min^{-1}). During the irradiation, the mixture was stirred at a constant speed. All experiments were performed at the natural pH ($\sim$ 3.5), except when studying the influence of the pH on the photocatalytic degradation of the substrate. In the investigation of the influence of electron acceptors, apart from constant streaming of O$_2$, H$_2$O$_2$, KBrO$_3$ or (NH$_4$)$_2$S$_2$O$_8$ was added to the CLP solution to make a 3 mM concentration. Where applicable, ethanol (400 μL) was added as a hydroxyl radical scavenger.

2.3 Analytical procedures

2.3.1 Kinetic studies

For the LC–DAD kinetic studies of the CLP photodegradation, samples of 0.50 mL of the reaction mixture were taken at the beginning of the experiment and at regular time intervals. Aliquot sampling caused a maximum volume variation of ca. 10% in the reaction mixture. Each aliquot was diluted to 10.00 mL with DDW. The obtained suspensions were filtered through a Millipore (Millex-GV, 0.22 μm) membrane filter. The absence of the CLP adsorption on the filters was preliminarily checked. After that, a 20 μL sample was injected and analysed on an Agilent Technologies 1100 Series liquid chromatograph, equipped with a UV/vis DAD set at 225 nm (absorption maximum for CLP), and a Zorbax Eclipse XDB-

C18 (150 mm × 4.6 mm i.d., particle size 5 μm, 25 ºC) column. The mobile phase (flow rate 1 mL min⁻¹, pH 2.56) was a mixture of ACN and water (3:7, v/v), the water being acidified with 0.1% H_3PO_4. Reproducibility of repeated runs was around 5–10%.

The total organic carbon (TOC) analysis was performed on an Elementar Liqui TOC II according to Standard US EPA Method 9060A. In studying the influence of the initial pH on the photocatalytic degradation use was made of a combined glass electrode (pH-Electrode SenTix 20, WTW) connected to a pH-meter (pH/Cond 340i, WTW).

2.3.2 Identification of the reaction intermediates

For the LC–ESI–MS/MS evaluation of intermediates, a 1.0 mM of CLP solution was prepared. Aliquots were taken at the beginning of the experiment and at regular time intervals during the irradiation. Then, a 20 μL sample was injected and analysed on an Agilent Technologies 1200 series liquid chromatograph with Agilent Technologies 6410A series electrospray ionisation triple-quadrupole MS/MS. The mobile phase (flow rate 1.0 mL min⁻¹) consisted of 0.05% aqueous formic acid and MeOH (gradient: 0 min 30% MeOH, 10 min 100% MeOH, 12 min 100% MeOH, post time 3 min). Components were separated on an Agilent Technologies XDB-C18 column (50 mm × 4.6 mm i.d., particle size 1.8 μm) held at 50 °C; UV/vis signal of the eluate was monitored at 210 nm, 225 nm and 260 nm (bandwidth 16 nm for each); continuous spectrum in the range from 200 to 400 nm (2 nm step) was also recorded. The eluate was forwarded to the MS/MS instrument without flow splitting. Analytes were ionised using the electrospray ion source, with nitrogen as drying gas (temperature 350 °C, flow 10 L min⁻¹) and nebuliser gas (45 psi), and a capillary voltage of 4.0 kV. High-purity nitrogen was used as the collision gas. Full-scan mode (m/z range 100–800, scan time 100 ms, fragmentor voltage 100 V), using positive or negative polarity (depending on compound), was used to select precursor ion for CLP and each degradation product, as well as to examine isotopic peaks distribution (Table 2). Then, product ion scan MS^2 mode (fragmentor voltage 100 V, scan time 100 ms, collision energy 0–40 V in 10 V increments) was used for structure elucidation of each degradation product.

2.4 Cell growth activity

Cell lines. For the estimation of cell growth effects, the cell lines H-4-II-E (rat hepatoma) and MRC-5 (human fetal lung) were grown in RPMI 1640 (H-4-II-E) and DMEM medium (MRC-5) with 4.5% glucose, supplemented with 10% heat inactivated FCS, 100 IU mL⁻¹ of penicillin and 100 μg mL⁻¹ of streptomycin. Investigated cell lines grew attached to the surface. They were cultured in 25 mL flasks (Corning, New York, USA) at 37 °C in atmosphere of 5% CO_2 and 100% humidity, sub-cultured twice a week and a single cell suspension was obtained using 0.1% trypsin with 0.04% EDTA.

Samples and controls used in cell growth experiments. For the analysis of cell growth effects, serial dilutions in distilled water were used. Samples were filtered through a 0.22 μm micro filters (Sartorius) to obtain sterility. The final concentrations of CLP before beginning the irradiation as well as in the CLP solution that was not irradiated were in the range from 6.25 to 100 μM, i.e. the dilution was from 10 to 160. Solution of CLP, filtered suspension of TiO_2 Wackherr catalyst and Aspirin® were used as negative controls, while cytotoxic drugs doxorubicin and gemcitabine, as well as $HgCl_2$ were used as positive controls.

	Compound	t_R (min)	M_{MI} (g mol^{-1})	UV max. (nm)	Mode	Precursor ion m/z	V_{col} (V)	Product ions m/z, rel. abundance
1	CLP	1.16	191	222, 281	NI	190	0	190 (82), 146 (100)
							10	146 (100)
							20	146 (100)
2	3,6-Dichloro-4,5-dihydroxypyridine-2-carboxylic acid	0.81	223	216, 278	NI	222	0	222 (51), 178 (100)
							10	178 (100), 142 (48), 106 (11)
							20	178 (71), 142 (100), 106 (40)
							30	142 (29), 106 (100), 78 (61), 66 (50)
3	3,6-Dichloro-pyridin-2-ol*	2.31	163	235, 315	PI	164	0	164 (100)
							10	164 (100), 146 (5), 124 (31)
							20	164 (43), 146 (86), 128 (100), 110 (37), 100 (26), 73 (78)
							30	146 (37), 128 (10), 110 (93), 75 (9), 73 (100), 62 (6)
					NI	162	40	110 (79), 75 (34), 73 (100), 62 (11)
							0	162 (100)
							10	162 (100)
4	3,6-Dichloro hydroxypyridine-2-carboxylic acid*	1.39	207	204, 248, 295	PI	208	0	208 (3), 190 (100)
							10	190 (100), 162 (42)
							20	190 (14), 162 (100), 107 (25)
							30	162 (45), 134 (9), 107 (100), 98 (15)
							40	107 (100), 98 (15)
					NI	206	0	206 (100), 162 (36)
							10	206 (12), 162 (100), 90 (11)
							20	162 (100), 126 (12), 90 (95), 62 (11)
							30	162 (7), 90 (100), 62 (94)
5	6-Chloro-3-hydroxypyridine-2-carboxylic acid or 3-Chloro-6-hydroxypyridine-2-carboxylic acid	1.57	173	233, 307	NI	172	0	172 (100), 128 (9)
							10	172 (54), 128 (100)
							20	128 (100)

Table 2. MS/MS fragmentation data of CLP photodegradation intermediates (Part I).

Compound	t_R (min)	M_{MI} (g mol^{-1})	UV max. (nm)	Mode	Precursor ion m/z	V_{col} (V)	Product ions m/z, rel. abundance
6 3,6-Dichloro pyridinediol	1.65	179	~227sh, 310, ~350sh	PI	180	0	180 (100)
						10	180 (100), 162 (6), 144 (11), 88 (12)
						20	180 (69), 162 (52), 144 (29), 107 (50), 98 (7), 89 (12), 88 (100)
						30	162 (9), 107 (100), 89 (9), 88 (53)
						40	107 (100), 98 (10), 89 (15), 88 (39), 72 (7), 53 (7), 52 (13)
				NI	178	0	178 (100), 142 (16)
						10	178 (100), 142 (51), 78 (21)
						20	78 (100)
7 3,6-Dichloro hydroxypyridine-2-carboxylic acid*	0.85	207	220, 265, 312–320**	PI	208	0	208 (100), 190 (24)
						10	190 (100), 162 (57)
						20	190 (12), 162 (100), 134 (11), 107 (59)
						30	162 (18), 107 (100)
						40	134 (10), 107 (100), 98 (7), 83 (8)
				NI	206	0	206 (92), 162 (100)
						10	162 (100), 126 (23), 90 (7)
						20	162 (45), 126 (76), 98 (52), 90 (100)
						30	126 (5), 98 (75), 90 (100), 66 (8), 62 (49)

* previously identified using TiO$_2$ Degussa P25
** wide spectral band
M_{MI} – monoisotopic weight

Table 2. MS/MS fragmentation data of CLP photodegradation intermediates (Part II).

Comparative Assessment of the Photocatalytic Efficiency of TiO$_2$ Wackherr in the Removal of Clopyralid from
Various Types of Water

171

Sulphorhodamine B (SRB) assay. Cell lines were harvested and plated into 96–well microtiter plates (Sarstedt, Newton, USA) at a seeding density of 4 x 10^3 cells per well (Četojevic-Simin et al., 2011), in a volume of 180 μL, and preincubated in complete medium supplemented with 5% FCS, at 37 °C for 24 h. Serial dilutions and solvent were added (20 μL/well) to achieve the required final concentrations and control. Microplates were then incubated at 37 °C for additional 48 h. Cell growth was evaluated by the colorimetric SRB assay according to Skehan et al. (1990). Cells were fixed with 50% TCA (1 h, +4 °C), washed with distilled water (Wellwash 4, Labsystems; Helsinki, Finland) and stained with 0.4% SRB (30 min, room temperature). The plates were then washed with 1% acetic acid to remove unbound dye. Protein-bound dye was extracted with 10 mM TRIS base. Absorbance was measured on a microplate reader (Multiscan Ascent, Labsystems) at 540/620 nm. The effect on cell growth was expressed as a percent of the control, and calculated as: % Control = (At/Ac) x 100 (%), where At is the absorbance of the test sample and Ac is the absorbance of the control.

Statistical analysis. The results of cell growth activity were expressed as mean ± SD of two independent experiments, each performed in quadruplicate (n = 8). Differences between control and treated groups were evaluated using one-way analysis of variance at the significance level of $p < 0.05$ (Microsoft Office Excel 2003 software). IC$_{50}$ values were calculated using Calcusyn for Windows (Version 1.1.0.0.; Biosoft).

3. Results and discussion

3.1 Effects of the type of TiO$_2$

The photocatalytic activity of TiO$_2$ Wackherr was compared to that of the most often used Degussa P25 under UV and visible irradiation. As can be seen from Figure 1, practically no degradation was observed under the visible light irradiation, either in the presence or absence of TiO$_2$. The lack of CLP disappearance in the presence of TiO$_2$ under these conditions also allows the exclusion of a significant adsorption of CLP on the catalyst surface during the course of the irradiation. In contrast, significant CLP removal could be observed under UV, and the process involving TiO$_2$ Wackherr was slightly faster compared to that observed in the presence of Degussa P25. This insignificant acceleration of the degradation of CLP in the presence of TiO$_2$ Wackherr is noteworthy, considering that this TiO$_2$ specimen has much larger particles (average radii in solution are 3–4 times larger) than Degussa P25 and a surface area that is almost six times lower (Vione et al., 2005).

The direct photolysis of CLP was also checked under the adopted irradiation conditions, in the absence of a catalyst (Figure 1). It appears that CLP can be degraded by direct photolysis in the near UV region, but at a significantly lower rate compared to the photocatalytic process.

Under the relevant experimental conditions, the reaction followed a pseudo-first order kinetics. On the basis of the kinetic curves ln c (substrate concentration) vs. t, the values of the pseudo-first order rate constant k' were calculated. The degradation rate of CLP was calculated for all the investigated as the product $k' c_0$, where c_0 is the initial concentration of CLP.

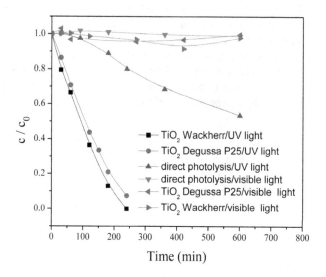

Fig. 1. Kinetics of the photolytic and photocatalytic degradation of CLP (1.0 mM). When present, the TiO$_2$ loading was 2.0 mg mL^{-1}. Operation conditions: $t = 40$ °C at pH ~ 3.5.

A comparison of the mineralisation capacities of TiO$_2$ Wackherr and Degussa P25, presented in Figure 2, shows that the mineralisation efficiency in the presence of TiO$_2$ Wackherr is significantly higher — 90% of CLP was mineralised during 240 min, whereas in the case of Degussa P25 only about 60%. Also, the ratio of the removal rate of the parent compound and total mineralisation in the case of TiO$_2$ Wackherr is about 1.20, whereas in the case of P25 this ratio is significantly higher, amounting to even 2.80. If these results are compared with those of the photocatalytic degradation of pyridine pesticides such as picloram and triclopyr (Abramović et al., 2011) it can be seen that the ratios of the mineralisation rate to the rate of the parent compound degradation are different and that they depend on the type of the substituent in the pyridine ring. Namely, in the photocatalytic degradation of triclopyr, TiO$_2$ Degussa P25 showed a higher efficiency both in the process of mineralisation and degradation of the parent compound. In the case of picloram, TiO$_2$ Wackherr showed higher photocatalytic efficiency than Degussa P25. However, the ratio of the rates of removal of the parent compound and total mineralisation in the presence of TiO$_2$ Wackherr is much higher in the case of picloram than of CLP, and it amounts to about 9, whereas in the presence of Degussa P25 this ratio is about 3, which is similar to the value obtained also for CLP.

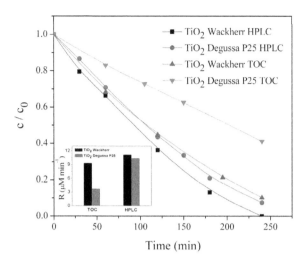

Fig. 2. Comparison of the photocatalytic removal parent compound and mineralisation of
CLP. The inset shows the degradation rate (R) calculated over 240 min of irradiation.
Operation conditions: $c(CLP)_0$ = 1.0 mM, TiO_2 = 2.0 mg mL^{-1}, t = 40 °C, at pH ~ 3.5.

3.2 Effect of catalyst loading

Due to the inherent nature of heterogeneous photocatalytic systems, there is always an
optimum catalyst concentration at which the removal rate is at its maximum. In this study,
the optimum was determined by changing the concentration of TiO_2 Wackherr over the
loading range from 0.2 to 2.0 mg mL^{-1}, as shown in Figure 3. As can be seen from the figure
inset, the increase in TiO_2 loading up to 1.0 mg mL^{-1} was accompanied by an increase in the
degradation rate, but a further increase caused an opposite effect. Theoretically, the increase
in the catalyst loading above an optimum value has no effect on the photodegradation rate
since all the light available is already utilized. However, higher loading of TiO_2 led to the
aggregation of its particles and thus to a decrease in the contact surface between the reactant
and photocatalyst particles, which caused a decrease in the number of active sites, resulting
in a lower rate of photodegradation. Also, when TiO_2 is overdosed, the intensity of the
incident UV light is attenuated because of the decreased light penetration and increased
scattering, which attenuates the positive effect coming from the dosage increment, and
therefore the overall performance decreases (Wong & Chu, 2003). Optimum catalyst
concentration is a complex function of a number of parameters including catalyst
agglomeration, the suspension opacity, light scattering, mixing, reactor type, and the
pollutant type (Toor et al., 2006; Mendez-Arriaga et al., 2008); hence it is not constant for all
photocatalytic systems. Indeed, optimum catalyst concentrations have been reported to vary
between as low as 0.1 mg mL^{-1} to as high as around 10 mg mL^{-1} (Mendez-Arriaga et al.,
2008; Alhakimi et al., 2008; Chu et al., 2009a). An optimum catalyst concentration of around
1 mg mL^{-1} has been generally reported in many studies (Chen & Ray, 1998; Lu et al., 1999;
Mendez-Arriaga et al., 2008; Rajeswari & Kanmani, 2009; Tizaoui et al., 2011). However, by
comparing these results with our previously finding that the optimal catalyst loading of

TiO$_2$ Degussa P25 was 4.0 mg mL^{-1} (Šojić et al., 2009), it can be concluded that the effect of catalyst loading on the efficiency of the photocatalytic removal of CLP is influenced by the type of TiO$_2$ used. If we compare the efficiencies of the two catalysts at their optimal loadings (i.e. 1.0 mg mL^{-1} for TiO$_2$ Wackherr and 4.0 mg mL^{-1} for Degussa P25), it comes out that TiO$_2$ Wackherr, although being present in a lower amount, is more efficient in the removal of CLP.

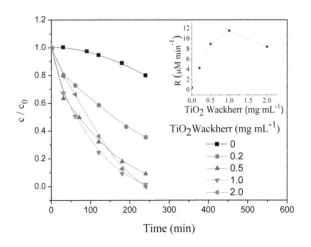

Fig. 3. Effect of the TiO$_2$ Wackherr catalyst loading on the kinetics of CLP photodegradation. The inset shows the effect of TiO$_2$ Wackherr loading on the degradation rate (R) determined after 120 min of irradiation. Operation conditions: $c(CLP)_0$ = 1.0 mM, t = 40 ºC at pH ~ 3.5.

3.3 Effect of the initial CLP concentration

Many studies have shown that the initial pollutant concentration has a significant effect on the rate of its photocatalytic removal. In this work, the effect of the initial concentration of CLP on the photodegradation rate was studied under UV light using TiO$_2$ Wackherr in the loading range from 0.25 to 1.0 mM (Figure 4). As can been seen from the inset of Figure 4, the degradation rate decreased with increase in the CLP concentration above to 0.5 mM. Such behaviour may be explained by the fact that at an increased concentration of CLP more of its molecules can be adsorbed on the photocatalyst surface, needing thus a larger catalyst area for their degradation. However, as the intensity of light, irradiation time and amount of catalyst are constant, the relative amounts of O$_2^-$ and ·OH radicals on the catalyst surface do not increase (Atiqur Rahman & Muneer, 2005; Qamar et al., 2006).

An alternative explanation for the effect of the substrate concentration is the competition for reactive species between the substrate and the transformation intermediates, the concentration of which increase with increasing substrate concentration, or the poisoning of the photocatalyst surface by the intermediates themselves (Abramović et al., 2011).

Comparative Assessment of the Photocatalytic Efficiency of TiO$_2$ Wackherr in the Removal of Clopyralid from Various Types of Water

175

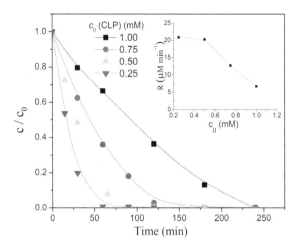

Fig. 4. Effect of the initial CLP concentration on the kinetics of photodegradation. The inset shows the effect of initial CLP concentration on the CLP degradation rate (R) calculated for 60 min of irradiation. Operation conditions: TiO$_2$ Wackherr = 2.0 mg mL^{-1}, t = 40 °C at pH ~ 3.5.

3.4 Effect of temperature

The photocatalytic degradation of CLP was studied in a temperatures range from 25 to 40 °C (Figure 5) and the rate constant k' was determined from the pseudo-first order plots. In general, the rate constant is expected to increase at higher temperatures, but it appeared that CLP more easily degraded at lower temperatures in the TiO$_2$ Wackherr suspension. Thus, the decrease in the rate constant observed in this temperature range may be attributed to the physisorption between the TiO$_2$ surface and the CLP molecules (Ishiki et al., 2005). Namely, the temperature of fastest CLP removal was, surprisingly, 25 °C, and hence all the further measurements were carried out at this temperature. The energy of activation, E_a, was calculated from the Arrhenius plot of ln$k$ versus $1/T$ (K^{-1}), and it amounted to 37.9 kJ mol^{-1}. Obviously, this value is somewhat higher than that obtained for the photocatalytic degradation of CLP in the presence TiO$_2$ Degussa P25 (Šojić et al., 2009), but it is acceptable since for TiO$_2$ photocatalyst, irradiation is the primary source of the electron-hole pair generation at ambient temperature, as the band gap energy is too high to be overcome by thermal activation (Topalov et al., 2004).

3.5 Effect of the initial pH

The effect of pH is very important in the heterogeneous photocatalytic removal of organic molecules since it influences both the surface charge of TiO$_2$ and the ionic form of the reactant, influencing thus the electrostatic interactions between the reactant species and the catalyst surface. Moreover, the pH influences the sizes of TiO$_2$ aggregates, interaction of the

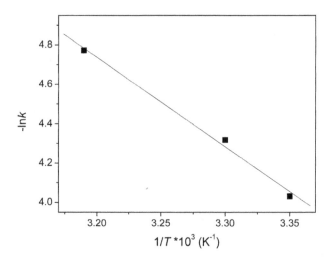

Fig. 5. Arrhenius plot of ln$k$ versus $1/T$ for the photocatalytic degradation of CLP for the first 120 min of irradiation. Operation conditions: $c(CLP)_0 = 1.0$ mM, TiO$_2$ Wackherr = 2.0 mg mL^{-1} at pH ~3.5.

solvent molecules with the catalyst and the type of radicals or intermediates formed during the photocatalytic reaction (Muneer & Bahnemann, 2002; Šojić at al., 2009; Tizaoui et al., 2011). All of these factors have a significant effect on the adsorption of solutes on TiO$_2$ surfaces and, as a result, on the observed photodegradation rates. Because the real effluent stuff of pesticide can be discharged at a different pH, the pH effect on the photocatalytic rate degradation of CLP was studied in the pH range from 2.4 to 9.8 (Figure 6). The point of zero charge (pH$_{pzc}$) of anatase is 5.8 (Karunakaran & Dhanalakshmi, 2009). Thus, the TiO$_2$ surface will be positively charged (TiOH$_2^+$) in acidic media (pH < pH$_{pzc}$) and negatively charged (TiO$^-$) in alkaline media (pH > pH$_{pzc}$). On the other hand, the pK_a values for CLP are 1.4±0.1 and 4.4±0.1 (Corredor et al., 2006), so that at pH < 1.4 the herbicide is mainly present in its protonated form, and at pH > 4.4 in the anionic form. In the pH interval from 2.4 to 3.5, one can expect a great increase in the photodegradation rate, arising as a consequence of the dissociation of the carboxylic group and deprotonation of the pyridine nitrogen (to a significantly smaller extent). In this way, favourable electrical forces are generated that are manifested as the attraction between the positively charged surface of the catalyst and CLP anion. As can be seen in Figure 6 (inset), in the pH interval from 3.5 to 4.8, a distinct decrease of the photodegradation rate is observed, arising probably as a consequence of the decrease in the number of positive sites on the catalyst surface. A further increase in the pH up to 9.8 caused a decrease in the photodegradation rate, which was probably a consequence of the influence of several factors. Namely, at pH > pH$_{pzc}$, the TiO$_2$ surface is negatively charged, causing the repulsion of the CLP anion. Besides, unfavourable electrical forces are generated, i.e., the repulsion between the negatively charged surface of the catalyst and OH$^-$.

Comparative Assessment of the Photocatalytic Efficiency of TiO₂ Wackherr in the Removal of Clopyralid from
Various Types of Water

177

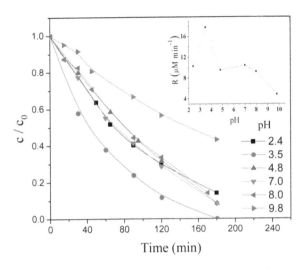

Fig. 6. Effect of the pH on the kinetics of CLP photocatalytic degradation. The inset shows
the effect of pH on the degradation rate (R) calculated for 120 min of irradiation. Operation
conditions: c (CLP)$_0$ = 1.0 mM, TiO₂ Wackherr = 2.0 mg mL^{-1}, t = 25 ºC.

3.6 Effect of electron acceptors

A practical problem arising in the use of TiO₂ as a photocatalyst is the undesired e⁻–h⁺ pair
recombination. One strategy to inhibit e⁻–h⁺ pair recombination is to add other (irreversible)
electron acceptors to the reaction mixture. They may have several different effects such as
(1) to increase the number of trapped electrons and, consequently, avoid recombination, (2)
to generate more radicals and other oxidising species, (3) to increase the oxidation rate of
intermediate compounds, and (4) to avoid problems caused by low oxygen concentration. In
highly toxic wastewater, where the degradation of organic pollutants is the major concern,
the addition of electron acceptors to enhance the degradation rate may often be justified
(Singh et al., 2007). The rates of photocatalytic degradation and mineralisation of CLP in the
presence of various electron acceptors such as KBrO₃, H₂O₂, and (NH₄)₂S₂O₈ in addition to
the molecular oxygen are shown in Figure 7.

As can be seen, the mentioned electron acceptors showed different effects. Namely, only the
addition of KBrO₃ enhanced the rate of photocatalytic degradation of the parent compound
(by a factor of 1.4), indicating that this compound is a more effective electron acceptor
compared with other oxidants employed in this study. A possible explanation might be the
change in the reaction mechanism of the photocatalytic degradation, since the reduction of
BrO₃⁻ by electrons does not lead directly to the formation of ˙OH, but rather to the
formation of other reactive radicals or oxidising reagents e.g. BrO₂⁻ and HOBr.
Furthermore, BrO₃⁻ by themselves can act as oxidising agents (Singh et al., 2007). However,
the mineralisation rate is slightly lower (by a factor of 1.1).

However, the presence of H_2O_2 caused a decrease in both the rate of removal of CLP (by a factor of 1.7) and its mineralisation (by a factor of 1.3). Such a negative effect of H_2O_2 is probably a consequence of the fact that it can also act as an $^{\bullet}OH$ scavenger, generating much less reactive hydroperoxyl radicals ($HO_2^{\bullet}$). The $HO_2^{\bullet}$ can further react with the remaining strong $^{\bullet}OH$ to form ineffective oxygen and water. Besides, at a higher dose, H_2O_2 might absorb and thus attenuate the incident UV light available for the photocatalysis process (Chu & Wong, 2004; Muruganandham & Swaminathan, 2006).

The presence of $S_2O_8^{2-}$ had an insignificant effect on the rate of degradation of CLP, but it decreased the rate of mineralisation more than $KBrO_3$ and H_2O_2, which can be explained by an increase in the concentration of SO_4^{2-} adsorbed on the TiO_2 surface, reducing thus the catalytic activity. The excess of adsorbed SO_4^{2-} also reacts with the photogenerated holes and with the $^{\bullet}OH$ (San et al., 2001; Muruganandham & Swaminathan, 2006).

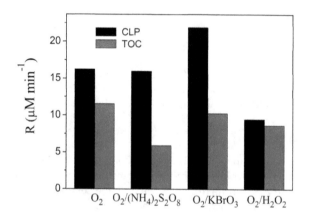

Fig. 7. Comparison of the degradation rate (R) of CLP removal and mineralisation in the presence of different electron acceptors (3 mM) calculated for 60 min of irradiation. Operation conditions: $c(CLP)_0 = 1.0$ mM, TiO_2 Wackherr = 2.0 mg mL^{-1}, $t = 25$ °C, pH ~3.5.

3.7 Effect of $^{\bullet}OH$ scavenger

In order to investigate whether the heterogeneous photocatalysis takes place via $^{\bullet}OH$, ethanol was added to the reaction mixture. Namely, it is known that alcohols, e.g. ethanol, act as $^{\bullet}OH$ scavengers (Daneshvar et al., 2004). The results obtained (data not shown) indicate that the degradation rate was significantly slower (by about 100 times) compared to that observed in the absence of ethanol, which proves that the reaction of photocatalytic degradation proceeded via $^{\bullet}OH$.

3.8 Intermediates and the mechanism of photodegradation

The LC-MS analysis of the irradiated CLP solutions indicated the formation of six intermediates (labelled 1–7, Table 2), whose kinetic curves are shown in Figure 8. Three of them, 3,6-dichloro-pyridin-2-ol (compound **3**) and isomeric 3,6-dichloro hydroxypyridine-2-carboxylic acids (compounds **4** and **7**) were previously identified (Šojić et al., 2009) in the presence of TiO₂ Degussa P25. Using the positive and negative ionization MS² spectra, it was possible to identify the remaining compounds and propose a photocatalytic degradation scheme (Figure 9).

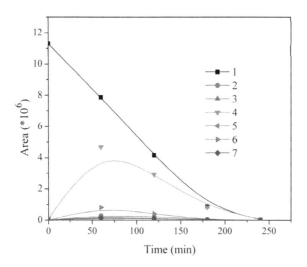

Fig. 8. Kinetics of the appearance/disappearance of CLP and intermediates in the photocatalytic degradation of CLP monitored by LC-ESI-MS/MS. Operation conditions: $c(CLP)_0 = 1.0$ mM, TiO₂ Wackherr = 2.0 mg mL⁻¹, $t = 25$ °C, pH ~3.5.

Compound **2**, eluting at 0.81 min, had M_{MI} 223, two chlorine atoms (on the basis of the A+2 isotopic peak intensity, as well as two consecutive losses of HCl in the MS² spectra: 178→142 and 142→106) and an odd number of nitrogen atoms (odd molecular weight), and was visible only in negative mode. In both the first-order and second-order MS spectra, the loss of CO_2 was observed (222→178), pointing out to the presence of carboxylic group. On the basis of the molecular weight (32 units higher than that of CLP) and spectral data, it was concluded that the compound is 3,6-dichloro-4,5-dihydroxypyridine-2-carboxylic acid.

Compound **5** eluted at 1.57 min. On the basis of the A+2 isotopic peak intensity and molecular weight, it could be concluded that it contains one chlorine atom and an odd number of nitrogen atoms. The only fragmentation observable in the NI MS¹ and MS² spectra was the loss of the carboxylic group as CO_2 (172→128). The monoisotopic weight of 173 mass units could be explained by the loss of one chlorine atom (which is in agreement with the isotopic profile) from the CLP molecule and introduction of one hydroxyl. Thus, the compound was identified as either 6-chloro-3-hydroxypyridine-2-carboxylic acid or 3-chloro-6-hydroxypyridine-2-carboxylic acid.

Finally, compound **6** was characterized by the odd monoisotopic weight of 179 units (pointing out to the odd number of nitrogen atoms), presence of two chlorine atoms, and the absence of carboxylic group loss both in positive mode (no sequential loss of H_2O and CO) and in negative mode (no loss of CO_2 or •COOH). Based on the molecular weight (12 units lower than that of CLP, which corresponds to the loss of COO and the introduction of two oxygen atoms), the compound was identified as 3,6-dichloro pyridinediol (the exact positions of hydroxyls could not be determined).

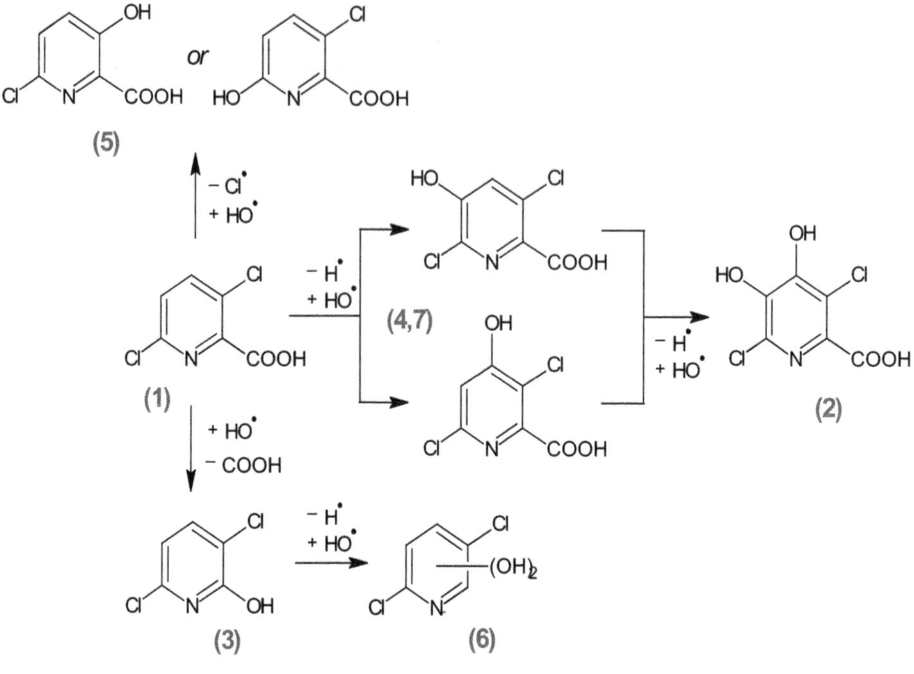

Fig. 9. Tentative pathways for photocatalytic degradation of CLP.

3.9 Cell growth activity

The cell growth activity of CLP, as well as of the mixture of CLP and its photocatalytic degradation intermediates was evaluated *in vitro* in a panel of two cell lines: H-4-II-E (rat hepatoma) and MRC-5 (human fetal lung) at 160, 80, 40, 20 and 10-fold dilutions that correspond to 6.25, 12.5, 25.0, 50.0 and 100 µM concentrations of CLP at the beginning of experiment (before the irradiation process). The toxicity was evaluated using the SRB assay (Skehan et al., 1990), which determines both specific rate of protein synthesis and cell growth rate.

Cell growth inhibition of CLP reached 5 to 7% in the MRC-5 and H-4-II-E cell line, respectively (Figure 10). The reaction mixture obtained after different irradiation times showed a higher toxicity toward the MRC-5 cell line compared to the parent compound after 120 min of irradiation at 20-fold dilution and after 240 min in the whole concentration range (Figure 10b). A comparison of the evolution of toxicity and degradation kinetics indicates that the toxicity toward the MRC-5 cell line was mildly increased after 120 min of irradiation at higher concentrations, i.e. at 20-fold dilution, and after 240 min in the whole concentration range. This implies that irradiation longer than 120 min contributed to the concentration of toxic degradation intermediates and to the toxicity of the mixture that is no longer dominated by the parent compound.

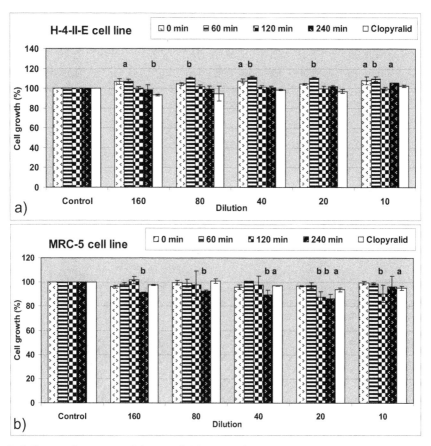

Fig. 10. Cell growth activity of the serial dilutions of CLP and its photocatalytic degradation intermediates obtained after different irradiation times in a) H-4-II-E and b) MRC-5 cell line. (One-way analysis of variance, compared to the control; **a**: $p < 0.05$, **b**: $p < 0.01$). Results are expressed as mean ± SD of two independent experiments, each performed in quadruplicate ($n = 8$).

On the other hand, in the H-4-II-E cell line, the reaction mixture obtained after 60 minutes of irradiation produced significant ($p < 0.01$) stimulation of cell growth compared to control (Figure 10a). The effects that are not concentration-dependent may be explained by the well known concept of hormesis (low dose stimulation and high dose inhibition). Since hormetic effects have been reported in a highly diverse array of biological models, for numerous organs and endpoints and chemical/physical stressors, it is evident that no single mechanism can account for these phenomena. In pharmacology, such dose responses have been studied with the aid of synthetic agonists and antagonists of receptors which mediate hormetic biphasic effects (Calabrese & Baldwin, 2001). A single agonist with differential binding (i.e. high and low receptor affinities) that affects two opposite acting receptors will induce hormetic-like biphasic dose responses in numerous biological systems, as has been shown for dozens of receptor systems. Pollutants, for example, may initiate significant changes in complex receptor systems, and affect biphasic dose responses by inducing changes in the concentrations of endogenous agonists. When such changes occur over a broad dose range, biphasic dose responses typically become manifested (Calabrese, 2005).

The IC_{50} values of Aspirin®, two well known cytotoxic drugs (Doxorubicin® and Gemcitabine®) and $HgCl_2$ (Table 3), as well as cell growth inhibition of TiO_2 Wackherr (Figure 11) were obtained in the same panel of cell lines.

Cell line	IC_{50} (μM)			
	ASP[a]	DOX[b]	GEM[c]	$HgCl_2$
H-4-II-E	>5551	0.272	0.004	3.189
MRC-5	>5551	0.408	0.384	69.578

[a]Aspirin®; [b]Doxorubicin®; [c]Gemcitabine®

Table 3. IC_{50} values (μM) of Aspirin®, Doxorubicin®, Gemcitabine® and $HgCl_2$ in selected cell lines.

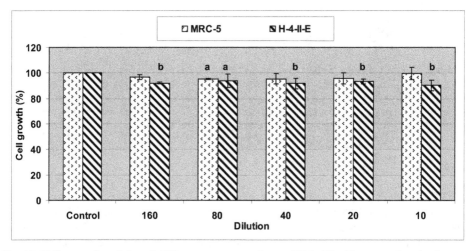

Fig. 11. Cell growth activity of different concentrations of TiO_2 Wackherr catalyst in the MRC-5 and H-4-II-E cell lines. (One-way analysis of variance, compared to the control; **a**: p<0.05, **b**: p < 0.01). Results are expressed as mean ± SD of two independent experiments, each performed in quadruplicate (n = 8).

In order to check whether the cell growth activity presented in Figure 10 is a consequence of the presence of CLP and its degradation intermediates only, it was necessary to run a blank test. To this end, aqueous suspension of TiO_2 Wackherr (2 mg cm[-1] without CPL) was sonicated in the dark for 15 min, as in the case of photodegradation of CLP, filtered through Millipore membrane filter, to apply then the same dilutions from 10 to 160. There were no significant effects (p < 0.01) on the growth of MRC-5 cell line. On the other hand, the

Comparative Assessment of the Photocatalytic Efficiency of TiO$_2$ Wackherr in the Removal of Clopyralid from Various Types of Water

183

H-4-II-E cell line appeared to be more sensitive to the presence of catalyst compared to the MRC-5 cell line (Figure 11). In the H-4-II-E cell line, the inhibition of cell growth influenced by TiO$_2$ Wackherr was significantly different compared to control ($p < 0.01$), even at the 160 dilution (Figure 11), but in all cases was below 10%. Solvent (DDW) was also tested after different irradiation times and it was shown to be nontoxic, i.e. all values were at the level of control that was treated with DDW (data not shown).

The effects of examined samples on the growth of selected cell lines were dependent on the type of cell line, concentration and time of irradiation. It can be concluded that CLP and reaction mixture of CLP and its photocatalytic degradation intermediates effected mildly the cell growth of both cell lines. In the examined concentration range, none of the treatments produced cell growth inhibition higher than 50%, i.e. the IC$_{50}$ values were not reached, either by the parent compound nor by samples obtained after different irradiation times. All examined samples exhibited lower toxicity in selected cell lines compared to the cytotoxicity of controls and HgCl$_2$.

A low level of free oxygen species is necessary for the promotion of cell proliferation (Burdon & Gill, 1993; Wei & Lee, 2002). The redox alterations play a significant role in a signal transduction pathway important for cell growth regulation. It is reasonable to propose that the examined samples obtained using UV irradiation in the presence of O$_2$ and TiO$_2$ Wackherr catalyst might influence the cell redox state, altering the cell proliferation.

Multi-endpoint bioassays that are based on whole cell response in human cell lines are a powerful indicator of metabolic, biochemical and genetic alterations that arise under the influence of evaluated compounds. This study presents an example of a systematic and simple first tier method to assess the toxicity of degradation products.

3.10 Effect of water type

Since natural aquatic systems contain dissolved organic matter (DOM) and different ionic species, it can be expected that they may complicate the photodegradation process. It has been reported that higher contents of inorganic and organic matter in tap and river water affect the efficiency of removal by UV/TiO$_2$ process (Buxton et al., 1988). To functionalise a TiO$_2$ water treatment process, the basic understanding of the effect of these inorganic ions on the photocatalytic performance is essential (Crittenden et al., 1996). Due to the zwitter ionic nature of the TiO$_2$ particles, it is also possible that the pH might have a profound effect on the selective inhibition of inorganic ions on the surface of the TiO$_2$ particles (Guillard et al., 2003). In this study, the effect of the matrix on the photocatalytic degradation of CLP was studied on the example of drinking and Danube water. As can be seen from Table 4, the rate of CLP removal from the samples of tap and river water was by about two/three times slower than that from DDW. The observed decrease in the degradation rate can be a consequence of the presence of HCO$_3^-$ and HUM in the examined water samples (Neppolian et al., 2002). Namely, the addition of HCO$_3^-$ and HUM to DDW in the amounts present in Danube water (Table 1), caused a decrease in the rate of photocatalytic degradation compared to that observed in DDW. As can been seen in Table 4, the concentration of HCO$_3^-$ in the examined tap water was somewhat higher, whereas the river water contained more HUM.

	HCO_3^- (mg L^{-1})	HUM (mg L^{-1})	R (µM min^{-1})
DDW			10.44
DDW	285	5	4.98
Tap water			5.41
DDW	182	15	4.51
Danube water			3.82

Table 4. The influence of water type on the degradation rate (R) of CLP determined after 120 min of irradiation. Operation conditions: $c(CLP)_0$ = 1.0 mM, TiO$_2$ Wackherr = 2.0 mg mL^{-1}, t = 25 ºC, pH ~7.0.

In the literature, the inhibition of photocatalytic properties in the presence of ions is often explained by the scavenging of •OH radical by ions. Of ionic species, HCO_3^- can especially, inhibit the degradation rate due to the high rate constant k' of its reaction with •OH (8.5 × 10^6 M^{-1} s^{-1}) (Buxton et al., 1988). Because of that we focused our attention on the influence of different concentrations of this ion on the photocatalytic degradation (Figure 12). Expectedly, an inhibition of CLP degradation was observed after adding HCO_3^- to DDW up to about 285 mg L^{-1}.

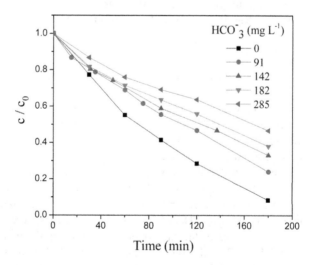

Fig. 12. Effect of the concentration of HCO_3^- on photodegradation of CLP in DDW. Operation conditions: $c(CLP)_0$ = 1.0 mM, TiO$_2$ Wackherr = 2.0 mg mL^{-1}, t = 25 ºC, pH~7.0.

The effect of HUM can be explained by the reaction with •OH, which lowers the availability of the latter for the reaction with CLP. Moreover, the actually available UV radiation reduces because some organic matters (especially aromatic compounds) absorb strongly UV irradiation (Chu et al., 2009b). Expectedly, the degradation rate decreased after the addition of HUM up to 20 mg L⁻¹ (Figure 13). The behaviour observed in the presence if HUM suggests a predominant effect of the •OH radical inhibition, due to the complex structure of HUM and their high reactivity with •OH radicals (Basfar et al., 2005; Prados-Joya et al., 2011).

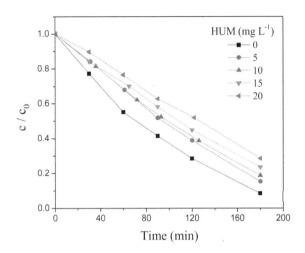

Fig. 13. Effect of the concentration of HUM on photodegradation of CLP in DDW. Operation conditions: $c(CLP)_0 = 1.0$ mM, TiO_2 Wackherr = 2.0 mg mL⁻¹, $t = 25$ °C, pH~7.0.

4. Conclusion

The results of this study clearly indicate that under the UV irradiation TiO₂ Wackherr was more efficient than Degussa P25 in both the process of removal of CLP from water and its mineralisation. The reaction followed the pseudo-first order kinetics. The optimum loading of TiO₂ Wackherr was 1.0 mg mL⁻¹ at pH 3.5. The photodegradation rate was dependent on the temperature, and the apparent activation energy was 37.9 kJ mol⁻¹. Along with molecular oxygen, KBrO₃ was the most efficient electron acceptor when concerning the degradation of the parent compound, whereas its mineralisation was most efficient in the presence of O₂ only. It was found that the presence of ethanol as a scavenger of •OH inhibited the CLP photodecomposition, suggesting that the reaction mechanism mainly involved free •OH. The LC–DAD, and LC–ESI–MS/MS monitoring of the process showed that six intermediates were formed. The analysis of the intermediate product formed during the photocatalytic degradation could be a useful source of information about the degradation pathways. The rate of photodegradation of CLP in DDW was about two/three times higher than in tap and river waters. The photodegradation rate was dominantly influenced by the pH of the medium and the presence of HCO_3^- and DOM. Our work validates the presented screening methodology of ecotoxicological risk assessment for transformation products, and can be used as a first step in toxicity assessment of degradation products and for prioritisation and planning of more detailed investigations.

5. Acknowledgment

This document has been produced with the financial assistance of the European Union (Project HU-SRB/0901/121/116 OCEEFPTRWR Optimization of Cost Effective and Environmentally Friendly Procedures for Treatment of Regional Water Resources). The contents of this document are the sole responsibility of the University of Novi Sad Faculty of Sciences and can under no circumstances be regarded as reflecting the position of the European Union and/or the Managing Authority and was supported by the Ministry of Education and Science of the Republic of Serbia (Projects: No ON172042).

6. References

Abramović, B.F. & Šojić, D.V. (2010). TiO$_2$-assisted Photocatalytic Degradation of Herbicides in Aqueous Solution. A review, In: *Desalination: Methods, Cost and Technoloqy*, Urboniene, I.A. (Eds.), pp. (17-142), Nova Science Publishers, Inc., ISBN 978-1-61668-909-4, New York

Abramović, B.F.; Anderluh, V.B.; Topalov, A.S. & Gaál, F.F. (2003). Direct Photolysis and Photocatalytic Degradation of 2-Amino-5-chloropyridine. *Journal of the Serbian Chemical Society*, Vol.68, No.12, (2003), pp. 961–970, ISSN 03525139

Abramović, B.F.; Anderluh, V.B.; Topalov, A.S. & Gaál, F.F. (2004a). Titanium Dioxide Mediated Photocatalytic Degradation of 3-Amino-2-chloropyridine. *Applied Catalysis B: Environmental*, Vol.48, No.3, (March 2004), pp. 213–221, ISSN 09263373

Abramović, B.F.; Anderluh, V.B.; Topalov, A.S. & Gaál, F.F. (2004b). Kinetics of Photocatalytic Removal of 2-Amino-5-chloropyridine from Water. *Acta Periodica Technologica*, Vol., No.35, (2004), pp. 79–86, ISSN 1450-7188

Abramović, B.F.; Anderluh, V.B.; Šojić, D.V. & Gaál, F.F. (2007). Photocatalytic Removal of the Herbicide Clopyralid from Water. *Journal of the Serbian Chemical Society*, Vol.72, No.12, (2007), pp. 1477–1486, ISSN 03525139

Abramović, B.F.; Banić, N.D. & Šojić, D.V. (2010). Degradation of Thiacloprid in Aqueous Solution by UV and UV/H$_2$O$_2$ Treatments. *Chemosphere*, Vol.81, No.1, (September 2010), pp. 114–119, ISSN 00456535

Abramović, B.; Šojić, D.; Despotović, V.; Vione, D.; Pazzi M. & Csanádi, J. (2011). A Comparative Study of the Activity of TiO$_2$ Wackherr and Degussa P25 in the Photocatalytic Degradation of Picloram. *Applied Catalysis B: Environmental*, Vol.105, No.1–2, (June 2011), pp. 191–198, ISSN 09263373

Alhakimi, G.; Studnicki, L.H. & Al-Ghazali, M. (2003). Photocatalytic Destruction of Potassium Hydrogen Phthalate Using TiO$_2$ and Sunlight: Application for the Treatment of Industrial Wastewater. *Journal of Photochemistry and Photobiology A: Chemistry*, Vol.154, No.2–3, (January 2003), pp. 219–228, ISSN 10106030

Atiqur Rahman, M. & Muneer, M. (2005). Heterogeneous Photocatalytic Degradation of Picloram, Dicamba, and Fluometuron in Aqueous Suspensions of Titanium Dioxide. *Journal of Environmental Science and Health-Part B Pesticides, Food Contaminants, and Agricultural Wastes*, Vol.40, No.2, (2005), pp. 247–267, ISSN 03601234

Banić, N.; Abramović, B.; Krstić, J.; Šojić, D.; Lončarević, D.; Cherkezova-Zheleva, Z. & Guzsvány, V. (2011). Photodegradation of Thiacloprid using Fe/TiO$_2$ as a

Heterogeneous Photo-Fenton Catalyst. *Applied Catalysis B: Environmental*, Vol.107, No.3–4, (September 2011), pp. 363–371, ISSN 09263373

Basfar, A.A.; Khan, H.M; Al-Shahrani, A.A. & Cooper, W.J. (2005). Radiation Induced Decomposition of Methyl Tert-butyl Ether in Water in Presence of Chloroform: Kinetic Modelling. *Water Research*, Vol.39, No.10, (May 2005), pp. 2085–2095, ISSN 00431354

Burdon, R.H. & Gill, V. (1993). Cellularly Generated Active Oxygen Species and HeLa Cell Proliferation. *Free Radical Research Communication*, Vol.19, No.3, (1993), pp. 203–213, ISSN 87550199

Buxton, G. V.; Greenstock, C. L.; Helman, W. P. & Ross, A. B. (1988). Critical Review of Rate Constants for Reactions of Hydrated Electrons, Hydrogen Atoms and Hydroxyl Radicals ($^\bullet OH/^\bullet O^-$) in Aqueous Solution. *Journal of Physical and Chemical Reference Data*, Vol.17, No.2, (June 1987), pp. 513–886, ISSN 0047-2689

Calabrese, E.J. (2005). Paradigm Lost, Paradigm Found: The re-emergence of Hormesis as a Fundamental Dose Reponse Model in the Toxicological Sciences. *Environmental Pollution*, Vol.138, No.3, (December 2005), pp. 379–412, ISSN 02697491

Calabrese, E.J. & Baldwin, L.A. (2001). Agonist Concentration Gradients as a Generalizable Regulatory Implementation Strategy. *Critical Reviews in Toxicology*, Vol.31, No.4–5, (2001), pp. 471–473, ISSN 10408444

Chen, D. & Ray, A.K. (1998). Photodegradation Kinetics of 4-Nitrophenol in TiO₂ Suspension. *Water Research*, Vol.32, No.11, (November 1998), pp. 3223–3234, ISSN 00431354

Chu, W. & Wong C.C. (2004). The Photocatalytic Degradation of Dicamba in TiO₂ Suspensions with the Help of Hydrogen Peroxide by Different Near UV Irradiations. *Water Research*, Vol.38, No.4, (February 2004), pp. 1037–1043, ISSN 00431354

Chu, W.; Rao, Y. & Hui, W.Y. (2009a). Removal of Simazine in UV/TiO₂ Heterogeneous System. *Journal of Agricultural and Food Chemistry*, Vol.57, No.15, (August 2009), pp. 6944–6949, ISSN 00218561

Chu, W.; Gao, N.; Li, C. & Cui, J. (2009b). Photochemical Degradation of Typical Halogenated Herbicide 2,4-D in Drinking Water with UV/H₂O₂/Micro-Aeration. *Science in China, Series B: Chemistry*, Vol.52, No.12, (December 2009), pp. 2351–2357, ISSN 10069291

Corredor, M.C.; Mellado, J.M.R. & Montoya, M.R. (2006). EC(EE) Process in the Reduction of the Herbicide Clopyralid on Mercury Electrodes. *Electrochimica Acta*, Vol.51, No.20 (May 2006), pp. 4302–4308, ISSN 00134686

Cox, C. (Winter 1998). Clopyralid, Herbicide Fact Sheet, In: *Journal of pesticide reform*, Vol.18 No.4, (access: 26. 8. 2011), Available from: http://www.mindfully.org/Pesticide/Clopyralid.htm

Crittenden, J.C.; Zhang, Y.; Hand, D.W.; Perram, D.L. & Marchand, E.G. (1996). Solar Detoxification of Fuel-contaminated Groundwater Using Fixed-Bed Photocatalysts. *Water Environmental Research*, Vol.68, No.3, (May 1996), pp. 270–278, ISSN 10614303

Četojević-Simin, D.D.; Velićanski, S.A.; Cvetković, D.D.; Markov, S.L.; Mrđanović, Z.J.; Bogdanović, V.V. & Šolajić V.S. (2010). Bioactivity of Lemon Balm Kombucha. *Food and Bioprocess Technology*, pp. 1–10, ISSN 19355130 (in press)

Daneshvar, N.; Salari, D. & Khataee, A.R. (2004). Photocatalytic Degradation of Azo Dye Acid Red 14 in Water on ZnO as an Alternative Catalyst to TiO$_2$. *Journal of Photochemistry and Photobiology A: Chemistry*, Vol.162, No.2-3, (March 2004), pp. 317-322, ISSN 10106030

Donald, D.B.; Cessna, A.J.; Sverko, E. & Glozier, N.E. (2007). Pesticides in Surface Drinking-Water Supplies of the Northern Great Plains. *Environmental Health Perspectives*, Vol.115, No.8, (August 2007), pp. 1183-1191, ISSN 00916765

Guillard, C.; Lachheb, H.; Houas, A.; Ksibi, M.; Elaloui, E. & Herrmann, J.-M. (2003). Influence of Chemical Structure of Dyes, of pH and of Inorganic Salts on Their Photocatalytic Degradation by TiO$_2$ Comparison of the Efficiency of Powder and Supported TiO$_2$, *Journal of Photochemistry and Photobiology A: Chemistry*, Vol.158, No.1, (May 2003), pp. 27-36, ISSN 10106030

Guzsvány, V.; Banić, N.; Papp, Z.; Gaál, F. & Abramović, B. (2010). Comparison of Different Iron-based Catalysts for Photocatalytic Removal of Imidacloprid. *Reaction Kinetics, Mechanisms and Catalysis*, Vol.99, No.1, (February 2010), pp. 225-233, ISSN 18785190

Huang, X.; Pedersen, T.; Fischer, M.; White, R. & Young, T.M. (2004). Herbicide Runoff along Highways. 1 Field Observations. *Environmental Science and Technology*, Vol.38, No.12, (May 2004), pp. 3263-3271, ISSN 0013-936X

Ishiki, R.R.; Ishiki, H.M. & Takashima, K. (2005). Photocatalytic Degradation of Imazethapyr Herbicide at TiO$_2$/H$_2$O Interface. *Chemosphere*, Vol.58, No.10, (March 2005), pp. 1461-1469, ISSN 00456535

Karunakaran, C. & Dhanalakshmi, R. (2009). Substituent effect on nano TiO$_2$- and ZnO-catalyzed Phenol Photodegradation Rates. *International Journal of Chemical Kinetics*, Vol.41, No.4, (April 2009), pp. 275-283, ISSN 05388066

Lu, M.-C.; Chen, J.-N. & Tu, M.-F. (1999). Photocatalytic Oxidation of Propoxur in Aqueous Titanium Dioxide Suspensions. *Journal of Environmental Science and Health-Part B Pesticides, Food Contaminants, and Agricultural Wastes*, Vol.34, No.5, (1999), pp. 859-872, ISSN 03601234

Méndez Arriaga, F.; Esplugas, S. & Giménez, J. (2008). Photocatalytic Degradation of Non-Steroidal Anti-Inflammatory Drugs with TiO$_2$ and Simulated Solar Irradiation. *Water Research*, Vol.42, No.3, (February 2008), pp. 585-594, ISSN 00431354

Muneer, M. & Bahnemann, D. (2002). Semiconductor-mediated Photocatalyzed Degradation of Two Selected Pesticide Derivatives, Terbacil and 2,4,5-Tribromoimidazole, in Aqueous Suspension. *Applied Catalysis B: Environmental*, Vol.36, No.2, (February 2002), pp. 95-111, ISSN 09263373

Muruganandham, M. & Swaminathan, M. (2006). Photocatalytic Decolourisation and Degradation of Reactive Orange 4 by TiO$_2$-UV Process. *Dyes and Pigments*, Vol.68, No.2-3, (2006), pp. 133-142, ISSN 01437208

Neppolian, B.; Choi, H.C.; Sakthivel, S.; Arabindoo, B. & Murugesan, V. (2002). Solar Light Induced and TiO$_2$ Assisted Degradation of Textile Dye Reactive Blue 4. *Chemosphere*, Vol.46, No.8, (2002), pp. 1173-1181, ISSN 00456535

Prados-Joya, G.; Sánchez-Polo, M.; Rivera-Utrilla, J. & Ferro-garcía, M. (2011). Photodegradation of the Antibiotics Nitroimidazoles in Aqueous Solution by Ultraviolet Radiation. *Water Research*, Vol.45, No.1, (January 2011), pp. 393-403, ISSN 00431354

Comparative Assessment of the Photocatalytic Efficiency of TiO$_2$ Wackherr in the Removal of Clopyralid from Various Types of Water

189

Qamar, M.; Muneer, M. & Bahnemann, D. (2006). Heterogeneous Photocatalysed Degradation of Two Selected Pesticide Derivatives, Triclopyr and Daminozid in Aqueous Suspensions of Titanium Dioxide. *Journal of Environmental Management*, Vol.80, No.2, (July 2006), pp. 99–106, ISSN 03014797

Rajeswari, R. & Kanmani, S. (2009). A Study on Degradation of Pesticide Wastewater by TiO$_2$ Photocatalysis. *Journal of Scientific and Industrial Research*, Vol.68, No.12, (December 2009), pp. 1063–1067, ISSN 00224456

Rossatto, V.; Picatonotto, T.; Vione, D. & Carlotti, M.E. (2003). Behavior of Some Rheological Modifiers Used in Cosmetics under Photocatalytic Conditions. *Journal of Dispersion Science and Technology*, Vol.24, No.2, (March 2003), pp. 259–271, ISSN 01932691

Sakaliene, O.; Papiernik, S.K.; Koskinen, W.C.; Kavoliunaite, I. & Brazenaiteih, J. (2009). Using Lysimeters to Evaluate the Relative Mobility and Plant Uptake of Four Herbicides in a Rye Production System. *Journal of Agricultural and Food Chemistry*, Vol.57, No.5, (March 2009), pp. 1975–1981, ISSN 00218561

San, N.; Hatipoğlu, A.; Koçtürk, G. & Çinar, Z. (2001). Prediction of Primary Intermediates and the Photodegradation Kinetics of 3-Aminophenol in Aqueous TiO$_2$ Suspensions. *Journal of Photochemistry and Photobiology A: Chemistry*, Vol.139, No.2–3, (March 2001), pp. 225–232, ISSN 10106030

Singh, H.K.; Saquib, M.; Haque, M.M.; Muneer, M. & Bahnemann, D.W. (2007). Titanium Dioxide Mediated Photocatalysed Degradation of Phenoxyacetic Acid and 2,4,5-Trichlorophenoxyacetic Acid, in Aqueous Suspensions. *Journal of Molecular Catalysis A: Chemical*, Vol.264, No.1–2, (March 2007), pp. 66–72, ISSN 13811169

Skehan, P.; Storeng, R.; Scudiero, D.; Monks, A.; McMahon, J.; Vistica, D.; Warren, J.T.; Bokesch, H.; Kenney, S. & Boyd, M. R. (1990). New Colorimetric Cytotoxicity Assay for Anticancer-Drug Screening. *Journal of the National Cancer Institute*, Vol.82, No.13, (1990), pp. 1107–1112, ISSN 00278874

Stapleton, D.R.; Emery, R.J.; Mantzavinos, D. & Papadaki, M. (2006). Photolytic Destruction of Halogenated Pyridines in Wastewaters. *Process Safety and Environmental Protection*, Vol.84, No.4B, (July 2006), pp. 313–316, ISSN 09575820

Šojić, D.V.; Anderluh, V.B.; Orčić, D.Z. & Abramović, B.F. (2009). Photodegradation of Clopyralid in TiO$_2$ Suspensions: Identification of Intermediates and Reaction Pathways. *Journal of Hazardous Materials*, Vol.168, No.1, (August 2009), pp. 94–101, ISSN 03043894

Šojić, D.; Despotović, V.; Abramović, B.; Todorova, N.; Giannakopoulou, T. & Trapalis, C. (2010a). Photocatalytic Degradation of Mecoprop and Clopyralid in Aqueous Suspensions of Nanostructured N-doped TiO$_2$. *Molecules*, Vol.15, No.5, (May 2010), pp. 2994–3009, ISSN 14203049

Šojić, D.V.; Despotović, V.N.; Abazović, N.D.; Čomor, M.I. & Abramović, B.F. (2010b). Photocatalytic Degradation of Selected Herbicides in Aqueous Suspensions of Doped Titania under Visible Light Irradiation. *Journal of Hazardous Materials*, Vol.179, No.1–3, (July 2010), pp. 49–56, ISSN 03043894

Tizaoui, C.; Mezughi, K. & Bickley, R. (2011). Heterogeneous Photocatalytic Removal of the Herbicide Clopyralid and its Comparison with UV/H$_2$O$_2$ and Ozone Oxidation Techniques. *Desalination*, Vol.273, No.1, (June 2011), pp. 197–204, ISSN 00119164

Toor, A.P.; Verma, A.; Jotshi, C.K.; Bajpai, P.K. & Singh, V. (2006). Photocatalytic Degradation of Direct Yellow 12 Dye Using UV/TiO$_2$ in a Shallow Pond Slurry Reactor. *Dye and Pigments*, Vol.68, No.1, (January 2006), pp. 53–60, ISSN 01437208

Topalov, A.S.; Šojić, D.V.; Molnár-Gábor, D.A.; Abramović, B.F. & Čomor, M.J. (2004). Photocatalytic Activity of Synthesized Nanosized TiO$_2$ towards the Degradation of Herbicide Mecoprop. *Applied Catalysis B: Environmental*, Vol.54, No.2, (December 2004), pp. 125–133, ISSN 09263373

Vione, D.; Minero, C.; Maurino, V.; Carlotti, M.E.; Picatonotto, T. & Pelizzetti, E. (2005). Degradation of Phenol and Benzoic Acid in the Presence of TiO$_2$-based Heterogeneous Photocatalyst. *Applied Catalysis B: Environmental*, Vol.58, No.1-2, (June 2005), pp. 79–88, ISSN 09263373

Wei, Y.-H. & Lee, H.-C. (2002). Oxidative Stress, Mitochondrial DNA Mutation, and Impairment of Antioxidant Enzymes in Aging. *Experimental Biology and Medicine*, Vol.227, No.9, (October 2002), pp. 671–682, ISSN 00379727

Wong, C.C. & Chu, W. (2003). The Direct Photolysis and Photocatalytic Degradation of Alachlor at Different TiO$_2$ and UV Sources. *Chemosphere*, Vol.50, No.8, (March 2003), pp. 981–987, ISSN 00456535

11

Row Crop Herbicide Drift Effects on Water Bodies and Aquaculture

Peter Wesley Perschbacher[1], Regina Edziyie[1] and Gerald M. Ludwig[2]
[1]University of Arkansas at Pine Bluff, Center for Aquaculture/Fisheries
[2]H.K.D. Stuttgart National Aquaculture Research Center
United States of America

1. Introduction

Aquatic ecosystems produce substantial amounts of aquatic products; including all new sources of seafood, from aquaculture. Level land with clay soils and the availability of water supplies makes riverine alluvial plains favorable areas for row crops and aquaculture. Aquaculture ponds are susceptible to impacts from row crop production through drift of herbicides. To assess these impacts we have conducted field research in replicated mesocosms filled with water and associated naturally-occurring communities from various pond ecosystems and subjected to expected levels of drift from all major aerially-applied herbicides currently in use. Rather than an organismal approach and LC_{50}'s, data indicates community-level approaches better approximate ecosystem impacts. Herbicide drift that affects phytoplankton adversely or in a stimulatory manner will similarly impact the ecosystem, as phytoplankton produce oxygen, take up ammonia and nitrite and provide food for zooplankton. Drift levels are below toxic levels to most other aquatic organisms, including fish (Spradley, 1991). Drift amounts reaching water bodies and ponds, including fish ponds, depend on many factors, but the cumulative range is most affected by the size of the water body. Thus, other than in direct overflight, larger catfish ponds (6-8 ha) have a drift range of 1-10% and smaller more recent designs of 4 ha, 5-20%. Even smaller ponds, used for fingerling production and baitfish production (0.8-2 ha), may receive drift amounts of up to 30% of the field rate.Herbicide drift may be expected to impact small water bodies through death or reduction in the photosynthetic rats of phytoplankton, which could reduce the supply of dissolved oxygen, inhibit removal of toxic nitrogenous wastes, and reduce production of zooplankton by reducing their food supply. These conditions could also result in death, disease, or lower growth rates of managed or cultured fishes. Triazine herbicides (atrazine and simazine), as well as amides (propanil), phenylureas (diuron), triazones, uraciles and phenolics, act through inhibition of photosystem II (PSII) of photosynthesis (Cobb, 1992). They are widely used in agriculture, since they provide a low-cost basal weed control (Jay et al., 1997). Using mesocosms and naturally-occurring plankton communities in a multi-day study provides better extrapolations to real environments than laboratory studies on a single species (Juettner et al., 1995), and possibly prevent overestimate of impacts (Macinnis-Ng and Ralph, 2002). The major drift source is aerial application, with an estimated 20 X higher drift deposition compared to application by ground spray booms (Hill et al., 1994).

2. Evaluation of 40 aerially-applied row crop herbicide effects on water bodies

Recent studies at the University of Arkansas at Pine Bluff (UAPB) have assessed the effects of herbicide drift from 40 herbicides used on adjacent soybean, rice, cotton, corn and winter wheat row crops to plankton and water quality in adjacent flood plain ponds (Perschbacher et al., 1997, 2002, 2008; Perschbacher and Ludwig, 2004). Herbicide drift may be expected to impact ponds through death or a reduction in the photosynthetic rate of phytoplankton, which could reduce the supply of dissolved oxygen, inhibit removal of toxic nitrogenous wastes, and reduce production of zooplankton by reducing the food supply (Waiser and Robarts, 1997). Aerial application has drift deposition 20 times higher compared to application by ground spray booms (Hill et al., 1994).%). The mode of action of herbicides impacting phytoplankton, is reversable inhibition of photosynthesis at photosystem II (PSII) and should not be species specific (Cobb, 1992; Solomon et al., 1996). Photosystem II inhibitors are widely used in agriculture, since they provide a low-cost basal weed control (Solomon et al., 1997; Jay et al., 1997).

2.1 Methods and materials

These studies were conducted to determine if aerially-applied herbicides would cause measurable plankton and water quality changes in outdoor pool mesocosms filled with water from a fish pond. Rates used encompassed the estimated range of drift and a field rate (full) equivalent to direct application. The studies were conducted at the UAPB Aquaculture Research Station at the approximate time of the year when the respective herbicides are applied. The experimental plankton mesocosms used were above ground, circular 500-L fiberglass tanks arranged in four rows on a cement pad. When filled, water depth of tanks was 0.7 m (slightly less than the average depth of most fish ponds) and there was no mud substrate. Water surface area of each tank was 0.78 m^2 and diameter was 1.0 m, similar to those used in a prior study of atrazine effects on plankton and water quality (Juettner et al., 1995). Tanks were filled immediately prior to herbicide application with water pumped from an adjacent 0.1-ha pond.

Herbicides were applied over the tank surfaces at one of three levels: field rates (equal to overspray) and high and low drift rates of 1/10 and 1/100th of this level, respectively (Perschbacher et al., 1997). A control, without herbicide addition, was the fourth treatment. Each herbicide was tested at the recommended application rate (Baldwin et al., 2000). Commercial formulations were used without addition of adjuvants or wetting agents. Approximately 30 ml of distilled water was used to dissolve the herbicide. Each treatment was replicated three times in randomly- assigned tanks. Tanks were flushed and air-dried between trials.

Each herbicide was added to the tanks at approximately 0900. A set of measurements was taken immediately prior to application and again 24, 48 and 72 hours after application. If effects were noted, sampling was continued approximately weekly until morning oxygen (DO) levels of drift treatments did not significantly different from the control (ie. recovery). Dissolved oxygen is the water quality parameter most sensitive to herbicide effects (Juettner et al., 1995) and most critical to fish culture. Water temperature, dissolved oxygen and pH were measured with a multiprobe meter (OI Analytical, College Station, TX). Total ammonia nitrogen (TAN) and nitrite-nitrogen were measured by Nessler and diazotization methods (Hach Co., Loveland, CO), respectively. Unionized ammonia (UIA) levels were calculated from measured temperature, TAN and pH. Chlorophyll *a*, corrected for pheophytin *a*, and a 2-h light and dark bottle estimation of phytoplankton net primary productivity (NPP) by the oxygen method followed APHA (2005), except for use of ethanol as the solvent for chlorophyll (Nusch, 1980). Major zooplankton group concentrations were also determined in each replicate in the following manner. Three, 1-l samples were obtained with a tube sampler that encompassed the entire water column. The samples were concentrated by being strained through a 70-um Wisconsin plankton net and then preserved in 70% isopropyl alcohol. Samples were identified and quantified by using a Sedgwick-Rafter cell and a microscope (Ludwig, 1993). Statistical analysis was by SAS statistical software package. ANOVA (after pretesting for normality) and LSD were used to test for significant differences ($P \leq 0.05$) among treatments for each day during each trial.

2.2 Results and discussion

Atrazine lowered NPP on d 2; and effects on ecosystems from field studies have been summarized as short-lived, with quick recovery at concentrations less than 50 ug/l (Solomon et al., 1996). Solomon et al. (1996) observed stimulation of chlorophyll *a* on d 7 post-application of 50 ug/L atrazine. This was also found by us at d 7 with propanil (Perschbacher et al. 1997, 2002). Edziyie (2004) noted drift from propanil affected fry ponds with $\leq$10 ug/l chlorophyll *a* less than culture ponds with levels of 50-85 ug/l of chlorophyll *a*, such as were present in this study. This may explain the reduced effects of atrazine, but is in need of further study. Carfentrazone drift rates resulted in significantly lower rotifer and nauplii numbers compared to control levels the day after application, but not on the second day. Reductions ranged from 5-30% of control numbers and could not be explained from chlorophyll *a* data or net primary productivity. Zooplankton from diuron and atrazine were noted greatly impacted (Table 2). Propanil at 1 and 10% drift rates did not result in significant effects, although full field rates did (Perschbacher et al., 2002). Further evaluation of propanil is considered in the section 3.

These studies indicate that drift effects from 40 common aerially-applied herbicide applications on plankton and water quality were limited to atrazine, diuron and carfentrazone (Table 1). Of the 40 herbicides, diuron presents the greater risk for reduced water quality and for a longer time period, of at least 4 weeks (Table 2).

3. Evaluation of drift levels to small alluvial plain water bodies: atrazine, propanil and diuron

Small water bodies, equal to or less than 1.2 ha, in alluvial plains may be subjected to greater drift concentrations from adjacent row crops, due to reduced surface areas and volumes (Perschbacher and Ludwig 2007). These small ponds may be used for growing early and vulnerable stages of commercial aquaculture crops, and for fish consumption by farm pond owners. The three herbicides causing appreciable impacts, atrazine, propanil and diuron, were further tested at maximum drift rates expected of 30% of field rates.

Common Name	Trade Name	Date Applied	A.I (kg/ha)	Chl. α (ug/l)
Soybean				
Bentazon	Basagran	8/23	0.57	200
Imazaquin	Image	8/2	0.14	240
Fomesafen	Flexstar	8/16	0.43	250
Aciflourfen	Blazer	8/9	0.43	270
Fluzifop	Fusilade	8/16	0.10	240
Clethodim	Prism	7/26	0.07	300
Chlorimuron	Canopy	8/9	0.004	125
Glyphosphate	Roundup	8/2	0.43	500
Flumiclorac	Resource	6/8	0.045	135
Sethoxydim	Vantage	6/1	0.45	239
Carfentrazone*	Aim	3/2	0.03	400
Rice				
Clomazone	Command	5/23	0.60	280
Thiobencarb	Bolero	5/29	3.40	400
Pendamethalin	Prowl	6/5	1.10	250
Propanil	Stam	6/12	4.50	160
Quinclorac	Facet	6/20	0.60	450
Halosulfuron	Permit	6/27	0.07	475
Bensulfuron methyl	Londax	7/5	0.07	240
2.4-D-amine	2,4-D	7/5	1.70	45
Molinate	Ordram	7/25	5.60	450
Triclopyr	Grandstand	7/11	0.40	115
Fenoxyprop-ethyl	Acclaim	6/15	0.13	114
Cyhalofop	Clincher	7/5	0.30	65
Bispyribac-sodium	Regiment	7/12	0.036	114
Cotton				
Diuron (burndown)*	Direx	3/5	1.40	390
Diuron (defoliant)	Direx	9/23	0.165	850
Paraquat	Gramaxone	4/10	0.83	160
Quizalofop	Assure	6/18	0.05	300
Dimethipin	Harvade	9/16	0.15	750
Tribufos	Def	10/7	1.00	1075
Ethephon	Finish	10/14	1.76	1000
Sodium chlorate	Defol	10/21	5.30	520
Glufosinate	Liberty	3/13	0.55	344
Flumioxazin	Valor	4/6	0.03	334
Corn				
Mesotrione	Callisto	5/30	1.80	150
Metolachlor	Dual	3/8	0.10	350
Atrazine*	AAtrex	5/6	0.90	30
Rimsulfuron	TranXit	5/3	0.90	40
Nicosulfuron	Steadfast	4/29	0.90	105
Winter Wheat				
Thifensulfuron + Tribenuron	Harmony Extra	3/25	0.028	189
* significant effects noted				

Table 1. Summary of mesocosm tests of drift from aerially-applied herbicides by major crop, common name, trade name, date applied, recommended active ingredient (A.I.) field rate and approximate levels of pond plankton.

Days Post-Application	Diuron 1/100	Diuron 1/10	Atrazine 1/100	Atrazine 1/10
DO				
1	92*	92*	80	102
2	93*	81*	100	104
7	83*	71*	102	111*
Recovery (days)	21	>28	0	0
NPP				
1	49*	21*	124	105
2	37*	25*	82*	79*
7	41*	22*	84	82
Chlorophyll *a*				
1	97	95	110	102
2	99	96	89	93
7	83*	58*	96	115
pH				
1	96*	95*	100	100
2	98*	91*	100	100
7	89*	87*	100	100
TAN				
1	194*	122*	94	101
2	120	80	92	85
7	243*	356*	133	94
UIA				
1	100	50*	100	100
2	84	12*	80	87
7	35*	37*	150*	125
Nitrite-N				
1	100	100	112	93
2	120	20	141	151*
7	100	133	25	150
Rotifers				
1	67	78	200	150
2	126	226	128	57
7	115	96	22*	67
Copepod nauplii				
1	238	163	74	76
2	66	92	104	95
7	100	100	199	300
Copepod adults				
1	100	160	173	110
2	75	150	110	113
7	64	21*	120	94
Cladocerans*				
1	NA	NA	120	94
2	NA	NA	73	127
7	NA	NA	100	82
*no cladocerans observed in diuron trials				

DO = 0900 Dissolved Oxygen; Recovery = return of morning DO to control levels;
NPP = Net Primary Productivity; TAN = Total Ammonia Nitrogen; UIA = Unionized Ammonia

Table 2. Comparison of mean low (1/100 direct application rate) and high (1/10 direct application rate) drift effects of diuron and atrazine, expressed as percentage of control levels. Means significantly different ($P \leq 0.05$) from control means, indicated by *.

3.1 Methods and materials

The study was conducted at the University of Arkansas at Pine Bluff (UAPB) Aquaculture Research Station The experimental mesocosms were 500-l, above ground, circular fiberglass tanks arranged in four rows on a cement pad. When filled, water depth of the tanks was 0.7 m (slightly less than the average depth of most fish ponds) and there was no soil substrate. Surface area of each tank was 0.78 m² and diameter was 1.0 m, similar to those used by Juettner et al.(1995). Tanks were filled immediately prior to herbicide application with water pumped from an adjacent 0.1-ha Aquaculture Research Station experimental pond. Total dissolved solids were 290 mg/l, hardness 185 mg/l and alkalinity 197 mg/l as calcium carbonate.

Commercial formulations, without adjuvants or wetting agents, were applied over the tank surfaces at 30% of field rates (Baldwin et al., 2000) in four randomly selected pools each. Four additional pools received no herbicide and served as controls. The level used was equivalent to highest potential cumulative drift concentrations based on graphs in Hill et al. (1994) to water bodies of 1.2 ha surface area. The experimental dose was added to 30 ml of distilled water for more uniform application over the tank surface.

Immediately following filling, the first set of measurements were taken. The suite of measurements was subsequently taken 24, 48 and 72 h after application. If impacts were noted, sampling was continued approximately weekly until morning oxygen levels of drift treatments did not significantly differ from the control. Dissolved oxygen is the water quality parameter most sensitive to herbicide effects (Juettner et al., 1995) and most critical to aquatic life. Water temperature, dissolved oxygen, total dissolved solids (TDS), and pH were measured with a multiprobe meter (YSI, Yellow Springs, OH). Total ammonia nitrogen and nitrite-nitrogen were measured by Nessler and diazotization methods (Hach Co., Loveland, CO), respectively. Unionized ammonia levels were obtained from water temperature, TDS, TAN and pH. Chlorophyll a, corrected for pheophytin a and using ethanol as a solvent (Nusch, 1980), and a 2-h light and dark bottle estimation of net phytoplankton primary productivity by the oxygen method followed Standard Methods (APHA, 2005). Concentrations of the major zooplankton groups (rotifers, copepod nauplii, adult copepods and cladocerans) were also determined in each replicate in the following manner. Six, 1-L samples were obtained with a tube sampler that encompassed the entire water column. The samples were concentrated by being strained through a 70-um Wisconsin plankton net and then preserved in 70% isopropyl alcohol. Samples were identified and quantified by using a Sedgwick-Rafter cell and a microscope. Phytoplankton were enumerated and identified to genus (Prescott, 1962) in Sedgwick-Rafter cells with Whipple grid at 150X (APHA, 2005) from 20 ml unconcentrated samples obtained with a 0.9-m polyvinyl chloride (PVC) column sampler and preserved with 1 ml of formalin. Cyanobacteria were further identified to species using Cocke (1967). A randomized block design was used. Means from each sample date were tested for significant differences ($P \leq$ 0.05) with controls by paired, single tail Student's t-tests.

3.2 Results

Propanil levels were 58 ug/l and atrazine levels were 19.5 ug/l. Significant changes from control treatment values were found for several parameters in all three herbicide treatments (Perschbacher and Ludwig, 2007). Following application on 20 June, net primary productivity was significantly depressed on d 1 in the propanil treatments, but increased on d 2 and 3. Morning dissolved oxygen was lower on d 1-3, but not to critical levels. Also, in the presence of propanil, pH and consequently UIA were lower from d 1-3. Atrazine reduced morning DO on d 2 and 3, but not net primary productivity. Nitrite-N, however, was significantly higher on d 1. Phytoplankton total numbers, and the cyanobacterium *Chroococcus* sp. which dominated, were reduced by propanil on d 1-3; similarly affected by atrazine on d 2 and 3. Numbers of green algae, *Scenedesmus* sp. and *Coelastrum* sp., and diatoms were however stimulated by propanil and diatoms by atrazine. Zooplankton were little affected by either herbicide.

Due to the greater impacts of diuron (at levels equivalent to 30 ug/l), response of important environmental metrics to diuron drift are presented in Tables 3-5. No significant differences in pre-application sampling were found. Following application of diuron, net primary productivity was reduced by 97%, and recovered on d 7 (Table 1). Morning oxygen concentration also declined on day 1 by 32%, and was at stressful levels from d 2-3. Recovery was attained on d 14. Chlorophyll *a* and pheophytin *a* levels were significantly higher on d 2-14. Levels of pH were reduced by diuron addition from d 1-14. With lower pH values, unionized ammonia was significantly less from d 2-14. Plankton were also significantly impacted. Cyanobacteria, with the exception of *Chroococcus* spp., were reduced from d 1 and green algae, especially *Scenedesmus* spp. were stimulated (Table 4, 5). The other major group of phytoplankton, pinnate diatoms, were unchanged with the exception of a decline on d 7. In terms of percentage composition of the phytoplankton community, in diuron-treated mesocosms cyanobacteria declined from 24 to 20%, while green algae increased from 45 to 72%. Diatoms also declined from 26 to 8% (Table 5).

Zooplankton groups with significantly reduced mean abundances included: nauplii-616/l compared to control level of 1750/l on d 7, and cladocerans-0/l treatment level compared to 33/l on d 2. Copepod numbers however increased: from 1483/l control level to treatment level of 2133/l on d 3, and from 1150/l control level to the 2133/l treatment level on d 4. Rotifers were not impacted, in contrast to the findings of Zimba et al. (2002) who found an increase in rotifers.

3.3 Discussion

Diuron is a urea herbicide, that is 4-6 times more potent in photosynthesis inhibition than simazine herbicide (Ashton and Crafts, 1981). The concentration used in this study and representing the highest drift level of diuron (Direx) was 30 ug/l. Cyanobacteria were most susceptible to diuron, found previously with diuron (Zimba et al. 2002) and propanil and atrazine (Voronova and Pushkar 1985, Leboulanger et al. 2001, Perschbacher and Ludwig 2007). An increase in chlorophyll *a* was also noted by Ricart et al. (2009) in biofilms exposed to 0.07-9.0 ug/L diuron. This was attributed to a so-called "shade-adaptation" response to reduced photosynthetic efficiency from diuron. Zimba et al. (2002) observed no decrease in phytoplankton biomass, as measured by chlorophyll *a*, during 9 weekly treatments of 10 ug/l diuron each, but found the phytoplankton composition was altered. Numbers of filamentous cyanobacteria decreased, while ultraplankton coccoid cyanobacteria, diatoms and chlorophytes increased and chlorophyll *b* indicative of chlorophytes was significantly higher on one sample date.

Although drift levels of diuron in the present study were 3 times higher than in the Perschbacher and Ludwig (2004) study, which evaluated maximum drift effects to water bodies over 7 ha, inhibition of photosynthesis was longer lasting in the 2004 study. The dominance of cyanobacteria which formed surface scums and were thus unstable in the former study may have been responsible for the greater impacts, as found for propanil (Edziyie, 2004).

The present study found that in small eutrophic ponds, typical in agricultural environments and with relatively high chlorophyll *a* levels, short-term negative impacts would be expected on morning DO from atrazine, propanil and diuron. However, they may also benefit water quality by reducing pH, a major concern in eutrophic ponds utilized for recreational fish production and commercial fish culture (Barkoh et al., 2005; Ludwig et al., 2007) and which in turn resulted in lowered unionized ammonia levels.

Parameters	Treatment			Time	(d)		
		0	1	2	3	7	14
DO (mg/l)	C	16.13	14.83*	11.23*	8.63*	6.73*	8.73
	D	16.07	11.02	3.10	2.50	4.63	7.87
NPP (mg O₂/l/h)	C	1.28	0.63*	0.51*	ND	0.13	0.35
	D	1.47	0.05	0.07	ND	0.22	0.32
Chlorophyll *a* (ug/l)	C	202.4	113.0	81.0*	70.8*	37.1*	ND
	D	209.2	131.6	126.5	108.0	118.1	ND
Pheophytin *a* (ug/l)	C	31.4*	41.7	19.4*	23.6	4.2*	ND
	D	45.9	47.9	38.7	33.7	21.2	ND
pH	C	8.57	8.73*	8.63*	8.42*	8.20*	8.47*
	D	8.60	8.60	8.07	7.73	7.75	8.17
UIA (mg/l)	C	0.01	0.02	0.02*	0.01*	0.01*	0.10*
	D	0.02	0.02	0.01	0.00	0.00	0.00

DO = 0900 Dissolved Oxygen; NPP = Net Primary Productivity; UIA = Unionized Ammonia; ND = No Data

Table 3. Mean (SE) water quality differences in diuron (D) and control (C) treatments. Column means significantly different have different letters ($P \leq 0.05$).

Species/Genera	Treatment		Time	(d)	
		0	2	7	14
Scenedesmus spp.	C	30.6	19.3	10.0	0.3*
	D	19.0	32.7	20.9	12.7
Ankistrodesmus spp.	C	1.4	2.3	1.5	0.5*
	D	2.1	2.5	1.5	2.5
Coelastrum spp.	C	1.1	0.8	0.1*	0.0*
	D	1.2	1.1	1.1	0.3
Anabaena levanderi	C	0.3	0.3	1.3*	12.3*
	D	0.2	0.1	0.0	0.0
Anabaena circinalis	C	0.1	0.2	0.5*	3.5*
	D	0.1	0.1	0.0	0.1
Oscillatoria angustissma	C	7.1	11.3*	27.3*	27.9*
	D	3.9	5.6	18.7	0.1
Chroococcus dispersus	C	3.3	2.7	6.7	0.2
	D	8.0	1.3	3.9	4.5
Pinnate diatoms	C	10.7	8.8	16.5*	1.0
	D	10.4	6.8	4.7	1.8

Table 4. Mean phytoplankton (10^3 cells/ml) in diuron (D) and control (C) treatments. Column means significantly different have * ($P \leq 0.05$).

Groups	Treatment		Time	(d)	
		0	2	7	14
Cyanobacteria	C	23.1	31.8*	55.7*	95.6*
	D	24.1	12.5	14.3	20.0
Green	C	51.6	47.4*	18.8*	2.0*
	D	45.4	71.8	68.1	72.0
Diatom	C	22.5	20.6	25.5	20.0*
	D	25.7	15.7	17.5	7.8

Table 5. Mean % composition of major phytoplankton groups by natural units, with and without diuron addition, over time. Column means significantly different are * ($P \leq 0.05$).

4. Modifiying factors due to algal state from *in situ* mesocosm testing

Aquaculture ponds often have surface floating scums predominately composed of cyanobacteria. These scum-forming algae are common in eutrophic ponds, including aquaculture ponds, especially during the growing season with warm temperatures and high nutrient loadings. Cyanobacteria in a suface scum state are unstable and prone to sudden die-offs (Boyd et al., 1975). The objective of this study was to test the effect of propanil on a pond with algal scums.

4.1 Methods and materials

The experiment was conducted at the University of Arkansas at Pine Bluff mesocosm facility. A completely randomized design was used, with three replicates for each treatment in 12 mesocosms and with water of approximately 400 ug/l chlorophyll *a* from a goldfish pond. The treatments used were: a control with no propanil, 1%, 10% and 100% of the recommended field rates (0.45 kg/ha). Variables measured included: morning dissolved oxygen, pH, nitrite-nitrogen, total ammonia nitrogen, unionized ammonia , net primary productivity, chlorophyll *a* and phytoplankton composition. Methods followed Standard Methods (APHA 2005). Samples were taken before and after treatments were added.

4.2 Results and discussion

Microsystis and *Anabaena* dominated the phytoplankton and formed the surface scum. Significantly lower DO and net primary productivity resulted after application in the 10% and full treatment. However, recovery was noted after 48 h. Also lower was pH following application. Tan and UIA were higher on d 2.

In the earlier trials (Perschbacher et al., 1997, 2002) without surface scum algae, propanil at 10% drift resulted only in elevated chlorophyll *a*, but no significant differences were noted in chlorophyll *a* and phytoplankton composition in the present study. Thus, the significant negative impacts found in the present study were not expressed in previous studies, and the difference is attributed to the algal state.

Effects of propanil drift depended on the level of chlorophyll *a* found in the systems, in the study by Edziyie (2005). The greatest impact was on water quality in ponds with chlorophyll *a* levels 50-200 ug/l and lesser impacts below 20 and above 300 ug/l. Phytoplankton at high levels have been proposed to modify pesticide effects by sorption to the algae (Day and Kaushik, 1987; Waiser and Robarts, 1997; Stampfli et al., 2011).

5. Conclusions

5.1 Large pond (7 ha and larger) simulations

Of the 40 herbicide applications tested, significant effects from drift levels of 1 and 10% (the range possible in ponds equal to and larger than 7 ha) were noted for diuron (used for burndown) and atrazine. Diuron presents the greater risk for reduced water quality and for a longer time period (in excess of 4 wks). Atrazine effects are short-lived . Carfentrazone resulted in brief zooplankton reductions.

5.2 Surface scum algal populations simulations

Algal populations forming scums appear more susceptible to these drift levels. Propanil levels which did not result in reductions in water quality in mixed water column populations, resulted in adverse reactions equal to the direct overspray. The concentration of algae at the surface and the propensity for algae in this stage to be unstable (crash) are judged responsible.

5.3 Differing chlorophyll *a* level simulations

Effects of propanil and atrazine drift, and perhaps of other herbicides, depend on the level of chlorophyll *a* found in the systems. The greatest impact of propanil was on water quality in ponds with chlorophyll *a* levels 50-200 ug/l and lesser impacts below 20 and above 300 ug/l. Absorption by algae, and other factors, may be responsible.

5.4 Small pond (1.2 ha and smaller) simulations

Simulations in small ponds, equal to or less than 1.2 ha, used drift rates up to 30%. Although atrazine and propanil did not cause concern, diuron caused DO drops that were below 3 mg/l for several days and recovery was not noted until 14 days.

5.5 Beneficial aspects of herbicide drift

Beneficial effects of atrazine, propanil and diuron included reduction or elimination of cyanobacteria, and reduced pH (Ludwig et al., 2007) and thus reduced UIA. Clorophyll *a* levels were stimulated by propanil and atrazine.

6. Acknowledgements

Students Baendo Lihono, and Malisa Hodges, and Jason Brown from USDA/ARS SNARC ably assisted the study. Dr. J. Dulka and Dupont kindly supplied the Basis Gold and component herbicides. C. Guy Jr., N. Slaton, H. Thomforde, W. Johnson, J. Ross, R Scott, and J. Welch of the Cooperative Extension Service are thanked for providing information on herbicides and herbicide samples. Virginia Perschbacher kindly assisted in manuscript preparation. Funding was from State of Arkansas, USDA/CSREES Project ARX 05013 and Grant No. 2001-52101-11300 Initiative for Future Agriculture and Food Systems and Catfish Farmers of Arkansas Promotion Board Grant #6.

7. References

APHA (American Public Health Association), (2005). *Standard Methods for the Examination of Water and Wastewater. 21st Edition,* American Public Health Association, ISBN 97780875530475, Washington, D.C, USA.

Ashton, F.& Crafts, A. (1981). *Mode of Action of Herbicides*. Wiley, ISBN 9780471048473, New York, NY, USA.

Baldwin, F.; Boyd, J. & Smith, K. (2000). *Recommended Chemicals for Weed and Brush Control*. University of Arkansas Cooperative Extension Service, Fayettville, AR. Publication #MP44-12M-1-00RV.

Barkoh, A.; Hamby, S., Kurten, G.& Schlechte, J. (2005). Effects of rice bran, cottonseed meal and alfalfa meal on pH and zooplankton. *North American Journal of Aquaculture*. 67(3), 237-243, ISSN 15222055.

Boyd, C.; Prather, E..& Parks, R. (1975). Sudden mortality of a massive phytoplankton bloom. *Weed Science* 23(1), 61-67, ISSN 00431745.

Chailfour, A. & Juneau P. (2011). Temperature-dependent sensitivity of growth and photosynthesis of *Scenedesmus obliquus*, *Navicula pelliculosa* and two strains of *Microcystis aerugninosa* to the herbicide atrazine. *Aquatic Toxicology* 103,9-17, 00431745Cobb, A. (1992). *Herbicides and Plant Physiology*. Chapman and Hall, ISBN 9780412438607, London, UK.

Cocke, E. (1967). *The Myxophyceae of North Carolina*. Edwards Publ., ASIN B0007E4A7S, Ann Arbor, MI, USA.

Day, K. & Kaushik, N. (1987). The absorption of fenvalerate to laboratory glassware and the alga *Chlamydomonas reinhardtii*, and its effect on uptake of the pesticide by *Daphnia galeata mendotae*. *Aquatic Toxicology* 10, 131-142, ISSN 00431745 .

Edziyie, R. (2004). *The Effect of Propanil on Five Pond Plankton Communities*. M.S. Thesis, University of Arkansas at Pine Bluff, Pine Bluff, AR, USA.

Hill, I.; Travis, K & Ekoniak, P. (1994). Spray drift and runoff simulations of foliar-applied pyrethiods to aquatic mesocosms: Rates, frequencies, and methods. In: *Aquatic Mesocosm Studies in Ecological Risk Assessment*, R. Graney, Kennedy, J.& Rodgers, J. (Eds.), 201-249, Lewis Publishers, ISBN 9780873715926, Boca Raton, FL, USA.

Hodson, R. & Hayes, M. (1989). *Hybrid striped bass pond production of fingerlings*. Southern RegionalAquaculture Center Publication 302, Stoneville, MS, USA.

Jay, A.; Ducruet, J.-M., Duval, J.-C. & Pelletier, J. (1997). A high-sensitivity chlorophyll flourescence assay for monitoring herbicide inhibition of photosystem II in the chlorophyte *Selenastrum capricornutum*: Comparison with effect on cell growth. *Archives Hydrobiology* 140, 273-286.

Juettner, I.; Peither, A., Lay, J., Kettrup, A. & Ormerod, S. (1995). An outdoor mesocosm study to assess the ecotoxico-logical effects of atrazine on a natural plankton community. *Archives Environmental Contamination and Toxicology* 29,435-441, ISSN 00904341.

Leboulanger, C.; Rimet, F., Heme de Lacotte, M. & Berard, A. (2001). Effects of atrazine and nicosulfuron on freshwater algae. *Environment International* 26, 131-135, ISSN 01604120.

Ludwig, G.; Hobbs, M. & Perschbacher, P. (2007). Ammonia, pH, and plankton in sunshine bass nursery ponds: the effect of inorganic fertilizer or sodium bicarbonate. *North American Journal of Aquaculture*. 69, 80-89, ISSN 15222055.

Macinnis-Ng, C. & Ralph, P. (2002). Towards a more ecologically relevant assessment of heavy metals on the photosynthesis of the seagrass *Zostera capricorni*. *Marine Pollution Bulletin* 45, 100-106, ISSN 0025326X.

Nusch, E. (1980). Comparison of different methods for chlorophyll and phaeopigment determination. *Archives Hydrobiology* 14, 14-36.

Perschbacher, P. & Ludwig, G. (2004). Effects of diuron and other aerially applied cotton herbicides and defoliants on the plankton communities of aquaculture ponds. *Aquaculture* 233, 197-203, ISSN 00448486.

Perschbacher, P. & Ludwig, G. (2007). High drift effects of propanil and Basis Gold on the plankton communities and water quality of a prestocking sunshine bass, *Morone chrysops* X *M. saxatilis*, fry pond. *Journal of Applied Aquaculture* 19, 101-111, ISSN 10454438.

Perschbacher, P.; Ludwig, G. & Edziyie, R. (2008). Effects of atrazine drift on production pond plankton communities and water quality using experimental mesocosms. *Journal of World Aquaculture Society* 39, 126-130, ISSN 17497345.

Perschbacher, P; Ludwig, G. & Slaton, N. (2002). Effects of common aerially-applied rice herbicides on the plankton communities of aquaculture ponds. *Aquaculture* 214, 241-246, , ISSN 00448486..

Perschbacher, P.; Stone, N., Ludwig, G. & Guy, C. (1997). Evaluation of effects of common aerially applied soybean herbicides and propanil on the plankton communities of aquaculture ponds. *Aquaculture* 157, 117-122, ISSN 00448486.

Prescott, G. (1962). *Algae of the Western Great Lakes Area*. William C. Brown Publishers, ISBN 0697045528, Dubuque, IA, USA.

Ricart, M.; Barcelo, D., Geizinger, A., Guasch, H., Lopez de Alda, M., Romani, A.M., Vidal, G., Villagrasa, M. & Sabater, S. (2000). Effects of low concentrations of the phenylurea herbicide diuron on biofilm algae and bacteria. *Chemosphere* 76, 1302-1401, ISSN 00456535.

Solomon, K.; Baker, D., Richards, R., Dixon, K., Klaine, S., La Point, T., Kendall, R., Weisskopf, C., Giddings, J., Geisy, J., Hall, L. & Williams, W., (1996). Ecological risk assessment of atrazine in North American surface waters. *Environmental Toxicology and Chemistry* 15, 31-76, ISSN 07307268.

Spradley, J. (1991). *Toxicity of pesticides to fish*. Publication MP330. University of Arkansas Division of Agriculture Cooperative Extension Service, Fayetteville, AR, USA.

Stampfli, N.; Knillmann, S., Matthias, L. & Beketov, M. (2011). Environmental context determines community sensitivity of freshwater zooplankton to a pesticide. *Aquatic Toxicology* 104, 116-120, ISSN 00431745.

Voronova, L. & Pushkar, I. (1985). Evaluating the impacts of agricultural pesticides and herbicides on aquatic biota. In: P. Mehrle, Gray, R. & Kendall, R. (Eds.), 29-37, *Toxic Substances in the Aquatic Environment: An International Aspect*, Water Quality Section American Fisheries Society, ISBN 9780913235350, Betheseda, MD. USA.

Waiser, M. & Robarts, R. (1997). Impacts of a herbicide and fertilizers on the microbial community of a saline prairie lake.*Canadian Journal Fisheries and Aquatic Science*. 54, 320-329, ISSN 0706652X.

Permissions

The contributors of this book come from diverse backgrounds, making this book a truly international effort. This book will bring forth new frontiers with its revolutionizing research information and detailed analysis of the nascent developments around the world.

We would like to thank Prof. Dr. Mohammed Naguib Abd El-Ghany Hasaneen, for lending his expertise to make the book truly unique. He has played a crucial role in the development of this book. Without his invaluable contribution this book wouldn't have been possible. He has made vital efforts to compile up to date information on the varied aspects of this subject to make this book a valuable addition to the collection of many professionals and students.

This book was conceptualized with the vision of imparting up-to-date information and advanced data in this field. To ensure the same, a matchless editorial board was set up. Every individual on the board went through rigorous rounds of assessment to prove their worth. After which they invested a large part of their time researching and compiling the most relevant data for our readers. Conferences and sessions were held from time to time between the editorial board and the contributing authors to present the data in the most comprehensible form. The editorial team has worked tirelessly to provide valuable and valid information to help people across the globe.

Every chapter published in this book has been scrutinized by our experts. Their significance has been extensively debated. The topics covered herein carry significant findings which will fuel the growth of the discipline. They may even be implemented as practical applications or may be referred to as a beginning point for another development. Chapters in this book were first published by InTech; hereby published with permission under the Creative Commons Attribution License or equivalent.

The editorial board has been involved in producing this book since its inception. They have spent rigorous hours researching and exploring the diverse topics which have resulted in the successful publishing of this book. They have passed on their knowledge of decades through this book. To expedite this challenging task, the publisher supported the team at every step. A small team of assistant editors was also appointed to further simplify the editing procedure and attain best results for the readers.

Our editorial team has been hand-picked from every corner of the world. Their multi-ethnicity adds dynamic inputs to the discussions which result in innovative outcomes. These outcomes are then further discussed with the researchers and contributors who give their valuable feedback and opinion regarding the same. The feedback is then collaborated with the researches and they are edited in a comprehensive manner to aid the understanding of the subject.

Apart from the editorial board, the designing team has also invested a significant amount of their time in understanding the subject and creating the most relevant covers. They scrutinized every image to scout for the most suitable representation of the subject and create an appropriate cover for the book.

The publishing team has been involved in this book since its early stages. They were actively engaged in every process, be it collecting the data, connecting with the contributors or procuring relevant information. The team has been an ardent support to the editorial, designing and production team. Their endless efforts to recruit the best for this project, has resulted in the accomplishment of this book. They are a veteran in the field of academics and their pool of knowledge is as vast as their experience in printing. Their expertise and guidance has proved useful at every step. Their uncompromising quality standards have made this book an exceptional effort. Their encouragement from time to time has been an inspiration for everyone.

The publisher and the editorial board hope that this book will prove to be a valuable piece of knowledge for researchers, students, practitioners and scholars across the globe.

List of Contributors

Margot Schulz and Felix Martin Ritter
University of Bonn, Institute of Molecular Physiology and Biotechnology of Plants (IMBIO), Germany

Dieter Sicker
University of Leipzig, Institute of Organic Chemistry, Germany

František Baluška
Institute of Cellular and Molecular Botany (IZMB), Germany

Heinrich W. Scherer
Institut für Nutzpflanzenwissenschaften und Ressourcenschutz (INRES), Plant Nutrition, Germany

Tina Sablofski
University of Bonn, Institute of Molecular Physiology and Biotechnology of Plants (IMBIO), Germany
Institute of Cellular and Molecular Botany (IZMB), Germany

Helena Prosen
University of Ljubljana, Faculty of Chemistry and Chemical Technology, Ljubljana, Slovenia

Fabiana A. Lobo
UFOP - Universidade Federal de Ouro Preto, Brazil

Carina L. de Aguirre, Patrícia M.S. Souza, André H. Rosa and Leonardo F. Fraceto
UNESP – State University of São Paulo, Brazil

Renato Grillo and Nathalie F.S. de Melo
UNESP – State University of São Paulo, Brazil
Department of Environmental Engineering, Campus Sorocaba, SP, Brazil

S.A. Clay and D.D. Malo
South Dakota State University, Plant Science Dept., Brookings, South Dakota, USA

Hai-Bo Yu, Xue-Ming Cheng and Bin Li
State Key Laboratory of the Discovery and Development of Novel Pesticide, Shenyang Research Institute of Chemical Industry Co. Ltd., China

Pilar Sandín-España, Beatriz Sevilla-Morán, José Luis Alonso-Prados and Inés Santín-Montanyá
Instituto Nacional de Investigación y Tecnología Agraria y Alimentaria (INIA), Spain

Flávio Martins Garcia Blanco and Antonio Batista Filho
Instituto Biológico de São Paulo, Centro Experimental, Campinas, Brazil

Edivaldo Domingues Velini
Universidade Estadual Paulista - UNESP, Faculdade de Ciências Agronômicas, Botucatu, Brazil

Rafael Grossi Botelho and Valdemar Luiz Tornisielo
Laboratório de Ecotoxicologia, Centro de Energia Nuclear na Agricultura, Universidade de São Paulo – CENA/USP, Piracicaba, SP, Brazil

João Pedro Cury and José Barbosa dos Santos
Universidade Federal dos Vales do Jequitinhonha e Mucuri – UFVJM, Diamantina, MG, Brazil

Ramiro Lascano, Nacira Muñoz, Germán Robert, Victorio Trippi and Gastón Quero
Instituto de Fitopatologia y Fisiologia Vegetal (IFFIVE-INTA), Córdoba, Argentina
Cátedra de Fisiología Vegetal; Facultad de Ciencias Exactas, Físicas y Naturales; U.N., Córdoba, Argentina

Marianela Rodriguez and Mariana Melchiorre
Instituto de Fitopatologia y Fisiologia Vegetal (IFFIVE-INTA), Córdoba, Argentina

Biljana Abramović, Vesna Despotović, Daniela Šojić, Ljiljana Rajić and Dejan Orčić
Faculty of Sciences, Department of Chemistry, Biochemistry and Environmental Protection, Novi Sad, Serbia

Dragana Četojević-Simin
Oncology Institute of Vojvodina, Sremska Kamenica, Serbia

Peter Wesley Perschbacher and Regina Edziyie
University of Arkansas at Pine Bluff, Center for Aquaculture/Fisheries, United States of America

Gerald M. Ludwig
H.K.D. Stuttgart National Aquaculture Research Center, United States of America